Digital Halftoning

Digital Halftoning

Robert Ulichney

The MIT Press
Cambridge, Massachusetts
London, England

Composed by the author with the L$_A$T$_E$X document preparation system. Camera ready copy of the resulting T$_E$X data file was printed on an Alphatype CRS phototypesetter by the American Mathematical Society.

The body type is Computer Modern Roman 10 point.

T$_E$X is a trademark of the American Mathematical Society.
Alphatype CRS is a trademark of Alphatype.

Second printing, 1988

Printed and bound by Halliday Lithograph in the United States of America.

Library of Congress Cataloging-in-Publication Data

Ulichney, Robert.
  Digital halftoning.

  "Based on a Ph.D. thesis at M.I.T. in 1986 entitled 'Digital halftoning and the physical reconstruction function'"
  Bibliography: p.
  Includes index.
  1. Computer graphics.   2. Image processing—Digital techniques.   I. Title.
T385.U45   1987   006.6   87-3182
ISBN 0-262-21009-6

Contents

List of Figures

Preface

Digital halftoning, also referred to as spatial dithering, is the method of rendering the illusion of continuous-tone pictures on displays that are capable of producing only binary picture elements. This book addresses the problem of developing such algorithms that best match the specific parameters of any target display device, particularly for nonstandard grid geometries. There is an emphasis on analysis in the spatial frequency domain.

This book provides a comprehensive catalog of halftoning techniques, allowing the reader to easily evaluate the tradeoffs. Techniques are organized by computational complexity and according to the nature of the dots produced, dispersed or clustered. The point process of dispersed-dot ordered dither is generalized for both rectangular and hexagonal grids. Hexagonal ordered dither proves to be the solution for asymmetric rectangular grids.

The concept of blue noise—high frequency white noise—is introduced and found to have desirable properties for halftoning. Very efficient algorithms for dithering with blue noise are developed, based on perturbed error diffusion. The nature of dither patterns produced is extensively examined in the frequency domain. New metrics for analyzing the frequency content of periodic and aperiodic patterns for both rectangular and hexagonal grids are developed.

Generalized sampling grids are also examined in detail; presented is a new "aspect ratio immunity" argument in favor of hexagonal grids. While some techniques benefit from the use of hexagonal grids, others are found to be ideally suited for rectangular grids.

Since the pictures tell the story, several carefully selected digitally produced examples are included.

A list of the major symbols used throughout this book is organized in a Glossary starting on page 345. An explanation of the notational conventions used is also included there. For both the Glossary and the list of References that follow it, cross references are provided to page numbers in this book where a symbol is defined or a citation is made.

This book is intended for designers and implementers of computer displays, digital image processing engineers, and signal processing theorists. It can serve as a supplementary text for a graduate-level course in digital image processing. The work is based on a EECS Ph.D. thesis at M.I.T. in 1986 entitled "Digital Halftoning and the Physical Reconstruction Function".

Digital Halftoning

Chapter 1

Introduction

Along with text and graphics, images will become a generic data type in general purpose computer systems. This poses new problems for the system designer. To the user, displaying an image on any of a wide variety of devices must be as transparent as displaying ASCII documents. Office video displays with differing gray level capacity, laser printers, and home dot-matrix printers, all of various resolutions and aspect ratios must render a given image in a similar way.

A solution requires that associated with each device is a dedicated display preprocessor that transforms the digital image data to a form tailored to the characteristics peculiar to that device.

Digital halftoning, a key component of such a preprocessor, refers to any algorithmic process which creates the illusion of continuous-tone images from the judicious arrangement of binary picture elements. It is often called *spatial dithering*. This book addresses the problem of developing such algorithms that best match the specific parameters of any target display device, modeled as the Physical Reconstruction Function.

Outside of using photographic film and some thermal sensitive materials, there does not exist a practical method of producing true continuous-tone hard copy. Computer hard copy devices are almost exclusively binary in nature. While the video displays associated with workstations and terminals are certainly capable of true continuous-tone, they are often implemented with frame buffers that provide high spatial resolution rather that full gray scale capability. Such devices are designed for dot-matrix text and graphics; digital halftoning provides the mechanism to display images on them.

The literature is replete with approaches to this problem, but almost all of them implicitly assume target displays with nonoverlapping symmetrically spaced dots on a rectangular raster. Often the grids are not symmetric, especially in the case of low resolution displays. It is frequently the case that resolution can be easily increased in one direction and not the other due to different physical constraints. Also, a conventional rectangular display can be made hexagonal by simply introducing a half pixel offset on every other line. Generalized techniques for halftoning on such grids are introduced in this report.

Understanding the nature of the dither patterns created by various algorithms is enhanced by examining their representation in the frequency domain. New techniques for summarizing the frequency content of periodic and aperiodic patterns on both rectangular and hexagonal grids are also presented.

In describing the parameters of the Physical Reconstruction Function in the next chapter, a new aspect ratio immunity argument in favor of hexagonal grids is developed. While some halftoning schemes benefit from the use of hexagonal grids, it will be shown in Chapter 8 through revelations in the frequency domain, that a rectangular grid is preferred for others.

It should be noted that the word *halftone* originates from the photoengraving process used in printing and has a well defined meaning that is somewhat broadened here. (For a historical review and digital application of the traditional printer's halftone screen see section 5.1.) It is borrowed in a liberal way to describe the theme of this book.

1.1   Choice of Halftoning Techniques

This is a study of controlled noise.

Contouring is a well known noise form resulting from coarse amplitude quantizing; artificial contours or boundaries develop in slowly varying regions of pictures that are truncated to a limited number gray levels. An extreme example illustrated in Figure 1.1 is when the number of gray levels is limited to two. (A good rendition of the original images can be seen on pages 297 and 298.)

Table 1.1: Categorization of Halftoning Techniques.

Chapter	Type of Pattern	Computational Complexity (Type of operation)	Type of "Dot"
4 white noise	aperiodic	point	dispersed
5 ordered dither	periodic	point	clustered
6, 7 ordered dither	periodic	point	dispersed
8 blue noise	aperiodic	neighborhood	dispersed

Here, the two test pictures that will be used throughout this text demonstrate dependence on image content. While the perception of gray level is completely gone, much of the detail of the scanned image survives. But, because many of the details in the synthesized image fluctuate entirely above or below the threshold, much of its content is obliterated.

Roberts [58] was the first to point out that dither does not increase noise energy, but simply redistributes that induced by fixed quantization in a way which makes it less visible. In the frequency domain, all of the error in coarse quantizing a fixed gray level is in the zero frequency (dc) term. A dithered rendition should have an error-free zero frequency term with all of the error scattered in higher frequency components.

Table 1.1 categorically lists the chapters which address particular halftoning techniques. A gray level can be rendered by covering a small area with either a clustered or dispersed "dot". If a display device can successfully accommodate an isolated black or white pixel, then by far the preferred choice is dispersed-dot halftoning which maximizes the use of resolution. A clustered-dot halftone mimics the photoengraving process used in printing, where tiny pixels collectively comprise dots of various sizes.

There is a choice of computationally complexity that can be accepted. A point operation in image processing refers to any algorithm

Figure 1.1: Quantizing with a Fixed Threshold.
(a) Scanned Picture.
(Compare with the better rendering on page 297.)

Figure 1.1: Quantizing with a Fixed Threshold. (b) Synthesized Image. (Compare with the better rendering on page 298.)

which produces output for a given location based only on the single input pixel at that location, independent of its neighbors. For applications where the minimization of computation time and/or hardware is a premium, then a point operation is preferred. Of course, neighborhood operations generally produce higher quality results.

All of the methods listed in Table 1.1 are viable options with the exception of dithering with white noise, which is presented for heuristic reasons only. The concept of dithering with blue noise, introduced in Chapter 8, achieves the uncorrelated features of white noise without the low frequency artifacts.

The preferred choice of a dispersed-dot point operation is ordered dither. The solution for hexagonal grids included in Chapter 6 proves to be the solution for asymmetric rectangular grids in Chapter 7.

A summary of the major techniques presented in this book are pictorially shown in Figure 1.2.

All of these techniques can be used to augment a display with limited gray scale capability. To best examine the nature of patterns produced, the focus of this work will be on the worst case, that of binary displays. For devices that can display more that two levels of gray, a simple extrapolation is explained at the end of this book in section 9.3.

Several overviews of existing halftoning algorithms have been written. Among them are surveys be Allebach [6], Jarvis, *et al.* [36], Stoffel and Moreland [77], Stucki [79], and Roetling [62]. A collection of halftoning papers are included in a book by Stoffel [78].

1.2 Image Rendering Systems

The display of high quality images on displays with only two levels of gray by halftoning can only be successful when performed as a component of the total image rendering *system*. The elements of such a system are identified in Figure 1.3.

The Physical Reconstruction Function is a system model of a given binary display device. It takes as its input a binary discrete-space image, $I[\mathbf{n}]$, and produces the continuous-space visual image, $I(\mathbf{x})$. What happens in this step varies widely from device to device. It is here that the image data is given the physical dimensions of a grid geometry,

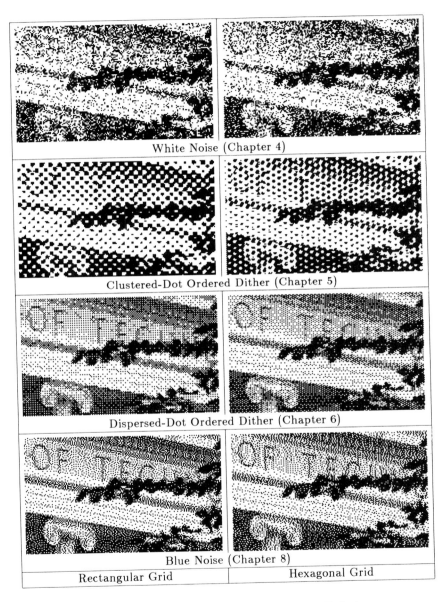

Figure 1.2: Pictorial Overview of Halftoning Techniques.

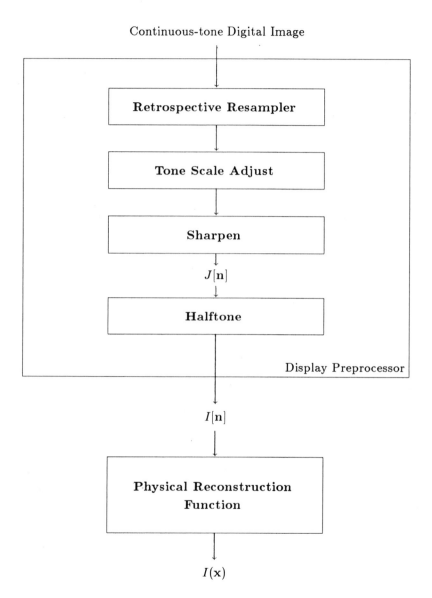

Figure 1.3: Image Rendering System.

with a "resolution" and aspect ratio. It is here that actual luminance values and dot structure are realized.

In a distributed system network, a given digital image may be displayed on any of several different hard copy and video devices. The role of the device dependent *display preprocessor* is to perform all the image processing manipulations necessary to transform a given continuous-tone digital image into a suitable intermediate binary digital image, $I[\mathbf{n}]$, that will yield the visible image when presented to the device.

The input to a halftoning process is a preprocessed continuous-tone digital image, $J[\mathbf{n}]$. Halftoning is the last step in the display preprocessor seen in Figure 1.3, after the other device dependent operations of Retrospective Resampler, Tone Scale Adjust, and optionally, Sharpen are executed.

The Retrospective Resampler is, in most cases, a digital scaler. Except for the special simple case of scaling by integral factors on a rectangular grid, some form of interpolation must be carried out. This is especially true if the given digital image and display device have different grid geometries, for example, one rectangular and the other hexagonal. The term *retrospective resampling*, adopted from an earlier work on scaling [82], describes the conceptual process of reconstructing the original continuous image from the given samples, then resampling this reconstruction.

A common method of performing such a reconstruction is through convolution with an appropriate interpolation function. Schreiber and Troxel have recently reported on the merits of such functions [69].

While resampling onto a new grid establishes the frame or "backbone" of the digital image, probably the most significant (and most underrated) contribution to image quality is tone scale. The simple point operation of mapping the gray level values of an image onto another distribution can have tremendous impact on the perceived quality of an image. The *Tone Scale Adjust* segment of the display preprocessor must compensate for the tone scale modification peculiar to the intended Physical Reconstruction Function in combination with the halftoning method to be used.

Once a halftoning algorithm has been selected for use on a particular device, a gray scale ramp should be generated on that device for the purpose of calibration. Physical measurements of the reflectance (for hardcopy devices) or luminance (for luminous devices) should be made of the output gray scale to determine the compensating tone scale table to be employed in the display preprocessor.

The next element seen in Figure 1.3 is the *Sharpen* operation. Sharpening is optional but if it is to occur it should take place before halftoning, usually in combination with resampling. At this point, a distinction should be made between image enhancement and the integrity of an image rendering system. In most cases digital images "look better" if some sharpening is performed. Such an operation can be classified as image enhancement, a process that creates a changed image which, by some criteria, is better than the original. The technique of creating the illusion of a gray level by the judicious distribution of binary pixels, the essence of halftoning, tends to unsharpen an image; fine image detail can be lost in halftone patterns. Presharpening an image to compensate for this effect, to maintain as closely as possible the unadulterated integrity of the original image, is not enhancement in the usual sense.

The virtues of a halftoning method should be evaluated in terms of its ability to render the illusion of gray scale with minimum visibility of algorithmic artifacts. Care must be used when evaluating and comparing methods which intrinsically sharpen. The "enhancement" perceived in the sharpened output can misleadingly outweigh other shortcomings in the algorithms. A precise degree of sharpening can be controlled independently of halftoning as will be shown in the last chapter.

1.3   Image Presentation Strategy

To assure a consistent and fair evaluation of all halftoning techniques to be presented, the same three specially selected source images are used throughout this text.

A historic picture of M.I.T. [51] of very high quality was digitized on both a rectangular and hexagonal grid with a constant number of samples per unit area. The image is a good test picture with textured and uniform regions, as well as areas of high detail such as the stone cut letters.

Scanned image data can sometimes benefit from noise inherent in the sampling process when halftoned. For that reason, a noiseless computer synthesized image borrowed from a work by Garcia [24] is occasionally shown for comparison.

While a halftoning scheme may perform well rendering the gray levels seen in a particular picture, it may fail on others. For this reason, a wrapped tone scale ramp revealing all gray levels is always shown for both rectangular and hexagonal grids. The ramp proceeds linearly from white to black marked at the beginning with a one pixel wide black line for reference.

All digitally generated images were laser recorded on RC paper by an ECRM Autokon[1] 8400, a device whose design and operation is described by Schreiber [65,67]. The device is capable of very high resolution output, but by means of pixel replication, the images in this text are displayed at very low resolution (about 33 dots/cm or 84 dots/inch for the case of a square grid). The reasons for this are to allow the reader to easily examine each of the dot patterns, and to assure that the images will survive reproduction. Higher resolution can be simulated by increasing the viewing distance. It should be noted that in the case of symmetric grids, the images in this text at the resolution shown would cover just over one square inch on a 300 dpi by 300 dpi display.

Several images are also shown on asymmetric rectangular grids, particularly in Chapter 7. To avoid overly small pixels for the same reasons stated above, the *smaller* dimension of any rectangular pixel is fixed to that of the square grids, at about 33 dots/cm. The other dimension will then have a lower resolution. Therefor, the overall resolution in pixels per unit area in this book is lower for images on asymmetric grids, by an amount proportional to the degree of asymmetry. For example, a picture shown on a grid with an aspect ratio of $\frac{1}{6}$ will have $\frac{1}{6}$ the number of pixels per unit area as that on a symmetric grid.

Further details about the grids used in this book are presented in section 2.1.2.2 (page 26).

1.3.1 Tone Scale Adjustment

The process of printing and reproducing this book can itself be modeled with a Physical Reconstruction Function. One important characteristic that must be compensated for is the darkening due to broadening of black pixels which darkens images.

---

[1]Autokon is a trademark of ECRM.

Figure 1.4 shows the actual tone scale adjustment curve used to prepare the rectangularly and hexagonally scanned images for the figures in this book. The transformation maps midrange gray levels to lighter values. (The straight superimposed line is a reference for a no-change transformation.)

Of particular importance are the horizontal portions at the top and bottom of the curve. They define the light and dark input ranges that are mapped to complete white and complete black. Such a tone scale clipping creates an effect referred to as "snap" or "punch" in the graphic arts to pictures that would otherwise be described as "flat". The adjustment is especially important for low resolution images as the ones in this book.

As an illustration of the dramatic difference a tone scale adjustment can make, Figure 1.5 is the scanned image used in all of the examples *without* the compensation of Figure 1.4. This should be compared to the identically halftoned picture on page 162.

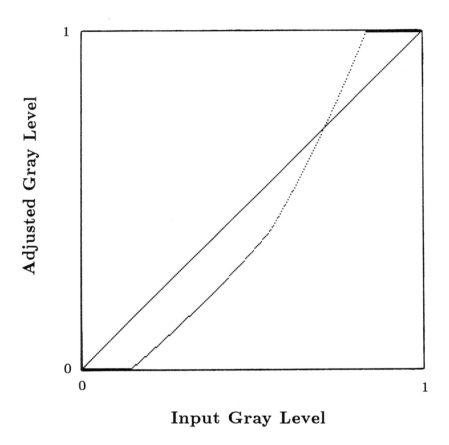

Figure 1.4: Tone Scale Adjustment used for the Scanned Picture.

Figure 1.5: "Scanned Picture" without Tone Scale Adjustment.
Compare with Figure 6.11(e) (page 162).

Chapter 2

Physical Reconstruction Function

Understanding the nature of binary displays is central to designing high quality halftoned images. The Physical Reconstruction Function, a general system model of a binary display shown in Figure 2.1, is the topic of this chapter. The only feature of this model which makes the display binary is the fact that it will accept only binary input, that is, the input digital image, $I[\mathbf{n}]$, is a discrete set of ones and zeros.

The first section will address the details of the first block, D/C or the Discrete-to-Continuous Space Converter, which is a mathematically convenient mechanism to map a set of numbers to physical two-dimensional space. It is in that section that a new argument in favor of hexagonal grids is made in terms of aspect ratio.

In section 2.2 attention will be paid to the linear shift invariant function, $d(\mathbf{x})$, which governs the nature of an individual output dot along with the generally nonlinear Tone Map, which assigns physical output luminance values to input values. In this model $d(\mathbf{x})$ is convolved with a two-dimensional array of impulses.

Noise

For completeness, two linear but space-varying components in the model of Figure 2.1 are included to describe the stochastic characteristics of

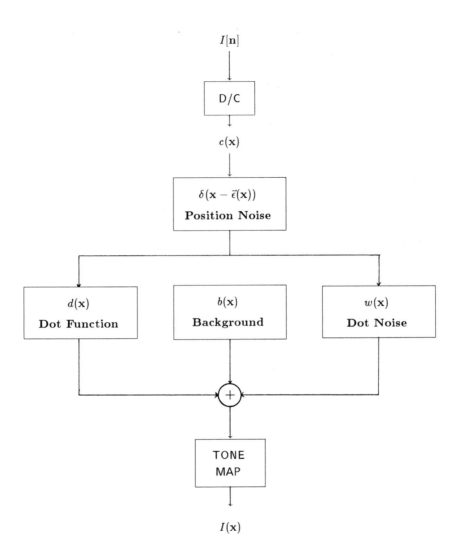

Figure 2.1: The Physical Reconstruction Function.

real physical devices.

Irregularities in the locations of dot centers is described by the Position Noise impulse. Dot-matrix impact printers have print wires that may wander in their solenoids. Perturbations in the dot positions of ink-jet printers are due to both aerodynamic and electrostatic interactions of drops in flight [40]. $\vec{\epsilon}(\mathbf{x})$ would usually be a zero mean random process, or could have a nonzero mean at spatially dependent locations, such as seen in column misalignments.

Ink spread, dot size fluctuations, and other local degradations are described by the Dot Noise, $w(\mathbf{x})$. The nature of this function can depend on anything from the size of the toner particles in laser printers [72] to the type of paper used [34].

Finally, to complete the Physical Reconstruction Function model, a "Background", $b(\mathbf{x})$, is added to describe the luminance from the paper or video phosphor used, along with any global degradations.

If the space varying, that is, the noise components are ignored, the relationship between the input and output of the Physical Reconstruction Function can be succinctly expressed as:

$$I(\mathbf{x}) = \textsf{TONE MAP} \left\{ \left(\sum_{\mathbf{n}} I[\mathbf{n}] \delta(\mathbf{x} - \mathbf{V}\mathbf{n}) \right) * d(\mathbf{x}) + b(\mathbf{x}) \right\}$$

where the sum is taken over all input locations, $\mathbf{n}$, "$*$" denotes convolution, and the matrix, $\mathbf{V}$, is described in the next section.

2.1   Grid Geometries

The content of this section is relevant not only to binary images and halftoning, but to digital image processing in general.

Perhaps the most important component of the Physical Reconstruction Function is the Discrete-to-Continuous Space Converter (D/C). It maps the input digital image, $I[\mathbf{n}]$, into a weighted set of delta functions in continuous space, $c(\mathbf{x})$, expressed in vector notation as

$$c(\mathbf{x}) = \sum_{\mathbf{n}} I[\mathbf{n}] \delta(\mathbf{x} - \mathbf{V}\mathbf{n}). \tag{2.1}$$

It is in this step that the set of numbers, $I[\mathbf{n}]$, takes on real dimensions; the Discrete-to-Continuous Space Converter establishes resolution and aspect ratio. The nature of the impulses in equation (2.1) is the topic of this section.

2.1.1 Periodic Sampling Grids

A *Periodic Sampling Grid* is a two dimensional impulse train,

$$\sum_{\mathbf{n}}^{\infty} \delta(\mathbf{x} - \mathbf{Vn}), \tag{2.2}$$

where the Sampling Matrix, $\mathbf{V} = [\mathbf{v}_1 \vdots \mathbf{v}_2]$, is composed of two linearly independent Sampling Vectors,

$$\mathbf{v}_1 = \left[\begin{array}{c} v_{11} \\ v_{21} \end{array} \right], \quad \mathbf{v}_2 = \left[\begin{array}{c} v_{12} \\ v_{22} \end{array} \right] \tag{2.3}$$

with reference coordinate system, $\mathbf{x}$, and index vector, $\mathbf{n}$.

Figure 2.2 shows a periodic sampling grid in its most general form. The sampling vectors, $\mathbf{v}_1$ and $\mathbf{v}_2$, can be thought of as grid generating vectors. Note that if they were not linearly independent, the sampling grid would be only one dimensional.

Since antiquity, the format (frame shape) used for the overwhelming majority of painted and printed pictures produced has been rectangular, rather than rhomboidal as suggested by the sampling vectors or any other shape. So, the reference coordinate system (x_1, x_2) should be orthogonal, as shown, regardless of the grid pattern generated. The reference system can, without loss of generality, be aligned anywhere. In this book, the orthogonal coordinate system (x_1', x_2') is adopted.

Image data is almost always organized in lines for digitizing, storing, and displaying. It is convenient to refer to the coordinates, x_1' and x_2', as the *sample* and *line* directions, respectively. Most often(but not always) the sample axis refers to the horizontal direction, and the line axis refers to the vertical dimension. The *sample period*, S, is the distance between grid points in the sample direction, and the *line period*, L, is the distance between lines. In terms of the sampling vectors, $S = |\mathbf{v}_1|$ and $L = |\mathbf{v}_2| \cos \theta$, where θ is the angle between $\mathbf{v}_1$ and $\mathbf{v}_2$.

Figure 2.2 also illustrates a natural way of defining pixel shape. *Pixel Shape* is defined as the smallest circumscribing polygon about a given

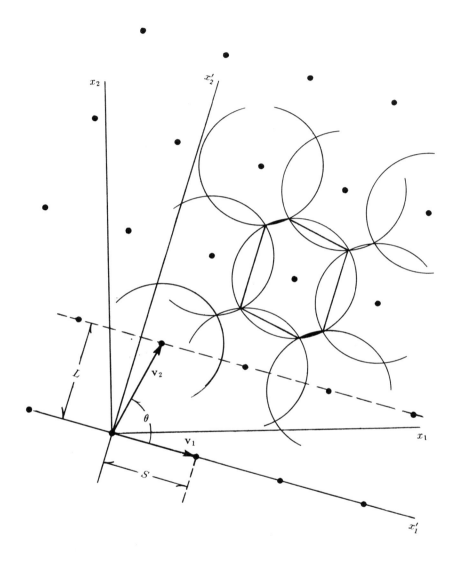

Figure 2.2: General Periodic Sampling Grid
with associated Pixel Shape.

grid point constructed from the perpendicular bisectors of lines between that point and all other grid points. Actual dot functions are most often circular in shape. The size of the circles is chosen to be just big enough to achieve complete coverage of the plane. For dots of this size and shape, an equivalent definition of pixel shape results: The polygon surrounding a grid point whose vertices are the intersections of circles of this size centered at each grid point. Note that for periodic sampling grids, pixels will always be either "hexagonal parallelograms" or rectangles. It should be noted that the area of such pixels is always $S \times L$ regardless of its shape.

2.1.2 Semiregular Grids

A general periodic sampling grid, that is, one with $\mathbf{v}_2$ uncontrained, has two shortcomings. Firstly, there is a lack of symmetry in the pixel shape and in the neighborhood surrounding it. Secondly, the offset for every line on an orthogonal coordinate system is different. This fact adds considerable complexity to simple image operations like cropping and scaling, or to the design of displays.

Semiregular Grids are defined as Periodic sampling grids whose corresponding pixel shapes are symmetric about at least two axes. There are two classes of semiregular grids:

1. *Rectangular Grids*, where $\mathbf{v}_1 \cdot \mathbf{v}_2 = 0$,   (that is, $\mathbf{v}_1 \perp \mathbf{v}_2$).

2. *Semiregular Hexagonal Grids*, where $\mathbf{v}_1 \cdot \mathbf{v}_2 = |\mathbf{v}_1|^2/2$,
   (or $|\mathbf{v}_2| \cos \theta = |\mathbf{v}_1|/2$).

Note that semiregular hexagonal grids require only one offset of exactly $S/2$ for every other line on an orthogonal coordinate system. Such grids have been referred to as "offset sampled" or "quincuncial". The familiar case of rectangular grids requires no offset.

An appropriate name for the shape of a pixel on semiregular hexagonal grids is *semiregular hexagon*. Of course, the shape of pixels on rectangular grids is rectangular. A useful metric which completely describes the shape of pixels on semiregular grids is aspect ratio. *Aspect Ratio* is defined as the ratio of the sample period, S, to the line period, L. That is, $\alpha \equiv S/L$.

2.1.2.1 Effect of Aspect Ratio

Aspect ratio is a parameter that often varies among display devices, particularly binary devices, but is seldom addressed theoretically. The effect of aspect ratio on pixel shape is shown for rectangular grid in Figure 2.3 and for semiregular hexagonal grids in Figure 2.4.

The shape of the pixel on a rectangular grid is a regular polygon, a square, for only one value of α ($\alpha = 1$). The shape of pixels on semiregular hexagonal grids is much more interesting. There are two cases where the pixel is a regular hexagon, for $\alpha = \frac{2}{\sqrt{3}}$ and $\alpha = 2\sqrt{3}$, and one special case where it is square, $\alpha = 2$.

To avoid confusion between the two kinds of hexagons, a *Hexagonal Grid of the First Kind* is defined to be a semiregular hexagonal grid with $\alpha < 2$ (see the top row of Figure 2.4). A *Hexagonal Grid of the Second Kind* is defined to be a semiregular hexagonal grid with $\alpha > 2$. (see the bottom row of Figure 2.4).

Mersereau [49,50] has shown that for a circularly band-limited waveform, sampling with a regular hexagonal grid involves 13.4% fewer samples to avoid aliasing (spectral overlap) than sampling with a square grid. This packing efficiency argument has long been recognized as one of the most important features of regular hexagonal grids. More recently, it has been shown [48] that hexagonal sampling produces samples with greater intersample dependency which allows "lost" samples to be more accurately recovered or restored. Another important argument can be made for semiregular hexagons.

In practical display devices, the physical constraints governing the size of the sample and line periods, S and L, are often very different. Accepting these values as fixed, a given rectangular pixel device can be converted into a hexagonal device by means of a very simple modification, introducing an alternate line offset of $S/2$. Assuming that the actual dots produced by the device are circles with a radius just large enough to achieve complete coverage of the image plane, a reasonable measure of performance is to compare the covering efficiency of the two arrangements. Note that in each case, rectangular and semiregular hexagonal, the aspect ratio and number of pixels per unit area remain constant.

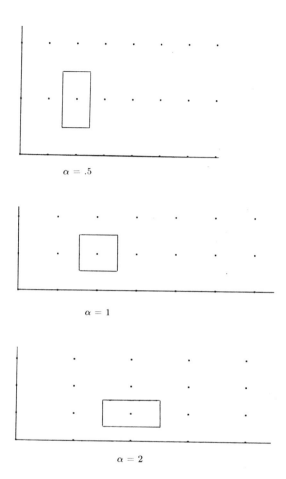

$\alpha = .5$

$\alpha = 1$

$\alpha = 2$

Figure 2.3: Pixel Shape on Rectangular Grids
as a function of Aspect Ratio, α.

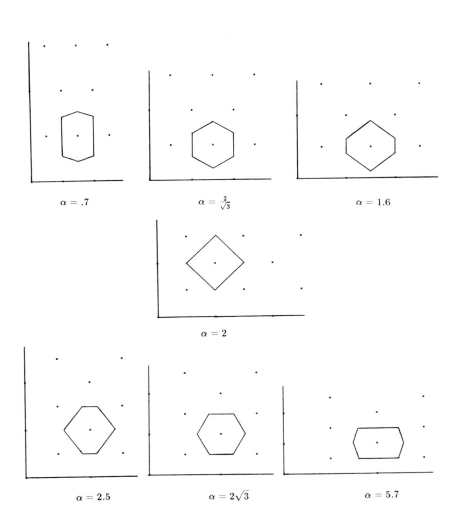

Figure 2.4: Pixel Shape on Hexagonal Grids
as a function of Aspect Ratio, α.

Covering Efficiency as a function of aspect ratio is defined as

$$E(\alpha) = \frac{\text{Pixel Area}}{\text{Circumscribing Circle Area}}.$$

A high covering efficiency is a desirable property for several reasons. Higher $E(\alpha)$ means

1. less dot overlap and thus a more linear tone scale rendition,

2. more similarly sized isolated black and white pixels, and

3. less spectral overlap (aliasing) for circularly band limited images.

As stated earlier, it can easily be shown that in all cases, pixel area is $S \times L$. The radius, r, of the circumscribing circle is equal to the distance from the pixel center to any of its vertices. Simple geometry reveals that

$$r = \begin{cases} \dfrac{\sqrt{S^2 + L^2}}{2} & \text{for rectangular grids.} \\[2ex] \dfrac{4L^2 + S^2}{8L} & \text{for hexagonal grids,} \quad \alpha \leq 2. \\[2ex] \dfrac{4L^2 + S^2}{4S} & \text{for hexagonal grids,} \quad \alpha \geq 2. \end{cases}$$

So, the resulting covering efficiency is

$$E(\alpha) = \begin{cases} \dfrac{4}{\pi(\alpha^{-1} + \alpha)} & \text{for rectangular grids.} \\[2ex] \dfrac{64\alpha^{-1}}{\pi(4\alpha^{-1} + \alpha)^2} & \text{for hexagonal grids,} \quad \alpha \leq 2. \\[2ex] \dfrac{16\alpha}{\pi(4\alpha^{-1} + \alpha)^2} & \text{for hexagonal grids,} \quad \alpha \geq 2. \end{cases} \qquad (2.4)$$

This function, plotted in Figure 2.5, reveals several interesting features. On a plot as this one where the abscissa, x, is equal to the logarithm of α, α is proportional to e^x and the rectangular function of equation (2.4) takes the form of a hyperbolic secant, as evidenced by its shape in Figure 2.5. The hexagonal curve is bimodal,

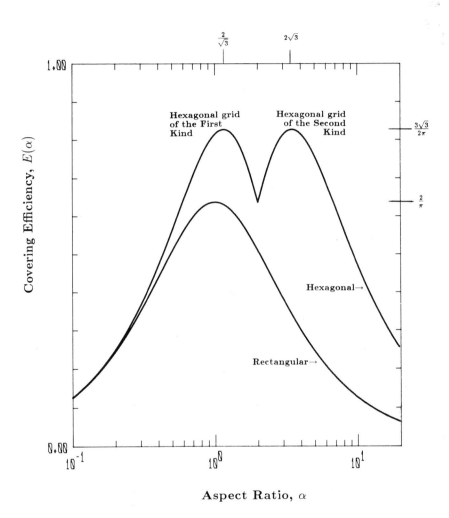

Figure 2.5: Covering Efficiency of Pixels on Semiregular Grids as a function of Aspect Ratio.

peaking at $E = \frac{3\sqrt{3}}{2\pi} \approx .827$ at the two aspect ratios where the pixel shapes are regular hexagons. At the cusp ($\alpha = 2$), where the hexagonal grid has square pixels, the covering efficiency, $E = \frac{2}{\pi} \approx .637$, is precisely that of the peak of the rectangular grid at $\alpha = 1$.

Probably the most important observation is that while semiregular hexagonal grids outperform rectangular grids at all aspect ratios, they achieve covering efficiencies better than or equal to the *best* rectangular case for aspect ratios between .591 and 6.77!

Digital halftoning employs binary displays where, particularly in the case of lower resolution devices, each pixel adds an appreciable contribution to the quality of the output. Along with the increased symmetry and decreased aliasing arguments, the relative "immunity" of hexagonal grids to aspect ratio make it an important alternative to rectangular grids.

2.1.2.2   Grids Used in this Book

Most of the digitally produced examples shown in this book are on square grids (rectangular grids with $\alpha = 1$) or regular hexagonal grids of the first kind (semiregular hexagonal grids with $\alpha = \frac{2}{\sqrt{3}}$), Unless otherwise specified, grids described as "rectangular" and "hexagonal" will refer to these cases.   To assure a fair comparison between these grids an equal number of samples per unit area is maintained. This is guaranteed by the equal-pixel-area condition, $S_r L_r = S_h L_h$, where the subscripts "r" and "h" are used to denote dimensions from rectangular and hexagonal grids. The rectangular examples are shown with $\alpha_r = S_r/L_r = 1$, and the hexagonal examples with $\alpha_h = S_h/L_h = \frac{2}{\sqrt{3}}$.

The rectangular and hexagonal sample periods, S_r and S_h, used as the unit of reference in the many frequency plots to follow, are thus related as

$$\frac{S_r}{S_h} = \sqrt{\frac{\sqrt{3}}{2}}.$$

These contraints leave one degree of freedom, the actual number of samples per unit area. For the case of regular grids, it has been set at 1096 pixels per square centimeter or roughly 33 pixels/cm (84 pixels/in) for the rectangular case. The reasons for this choice of low resolution were indicated in section 1.3.

Table 2.1: Some Major Display Classes

Class	$d(\mathbf{x})$	Tone Map	Examples
I	No Overlap	–	Liquid Crystal
			Electroluminescent
			Plasma Panel
II	Pillbox	Hard Step	Wire Impact (carbon ribbon)
III	Pillbox	Soft Step	Ink-jet
			Wire Impact
IV	Gaussian	Hard Step	Electrophotographic (Laser)
			Offset Printing
V	Gaussian	Soft Step	Cathode Ray Tube (Video)

2.2 Dot Function and Tone Map

The dot function, $d(\mathbf{x})$, is the linear space invariant component of the display with superposition described by convolution, while the generally nonlinear Tone Map assigns physical output luminances to input values. Next to the Discrete-to-Continuous Space Converter, it is these two components that most distinguish display devices.

Table 2.1 identifies some basic classes of binary displays. The five classes shown are not expected to be exhaustive, but exemplify target models for halftone display.

Most existing halftoning algorithms are implicitly designed for Class I displays (with square grids). This, the simplest class, includes any display with nonoverlapping dots. The resulting luminance depends only on the number of dots turned on, independent of any Tone Map. The other classes in table 2.1 all have overlapping dot functions. The Tone Maps for these classes are all described as "steps" since they eventually clip the output at some minimum and maximum. "Hard" steps have no transition region between the two extremes, and thus do not increase density at areas of overlap.

Photographic close-ups of a Class II device (a), and Class V devices (b and c) are shown in Figure 2.6. While the modeling of the Tone Map for the dot-matrix printer as a hard step is supported by the apparent lack of transition gray levels between the black of the dot and the white of the paper, a gradual transition (soft step) is seen in the video and laser recorder photographs.

A very important type of binary display is Class IV, which includes the popular electrophotographic (laser) printers. Plain paper electrophotographic printers operate by charging or discharging a photosensitive drum with a scanning laser illumination, leaving a latent image to which oppositely charged toner particles are attracted. Existing products are described as "positive" or "negative" printers depending on whether the laser erases white or writes black.

An example of how the dot function, $d(\mathbf{x})$, can itself be composed of several linear components cascaded together in convolution is shown qualitatively in Figure 2.7 for class IV and V devices. Sonnenberg [74,75] has carefully studied the tradeoffs that must be considered in setting the parameters for $d(\mathbf{x})$ in laser printers. To maintain the integrity of primitives occurring most often in text, thin horizontal and vertical lines are usually favored at the expense of isolated pixels, and thus dispersed-dot halftoning on such devices yield very nonlinear results.

Figure 2.8 illustrates both the nonlinearity and diversity in electrophotographic printer output. Since this is a cursory rather than thorough comparison of two products, the names "Product X" and "Product Y" are used. Both printers have the same resolution of 118 dots/cm (300 dots/inch). They, along with the ideal image of the data to be printed, are shown at the same magnification. The difference between Figure 2.8 (b) and (c) is due to a tighter dot size and smaller toner particles in (c); both printers are "positive" in that the laser erases white area. While "Product Y" is more successful in printing an isolated black dot, both display a disproportionate increase in density in regions where dots are clustered.

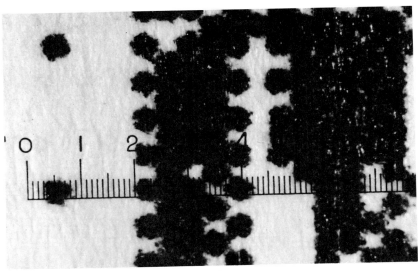

(a) 14× magnification of output from a 24 by 28 dots/cm dot-matrix printer.
(Numbered reticle marks are millimeters.)

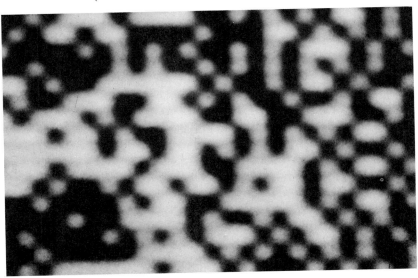

(b) 14× magnification of output from a 31 dots/cm Video Terminal.
(This photo is *not* out of focus.)

Figure 2.6: Photographic Enlargements of Binary Display Output.

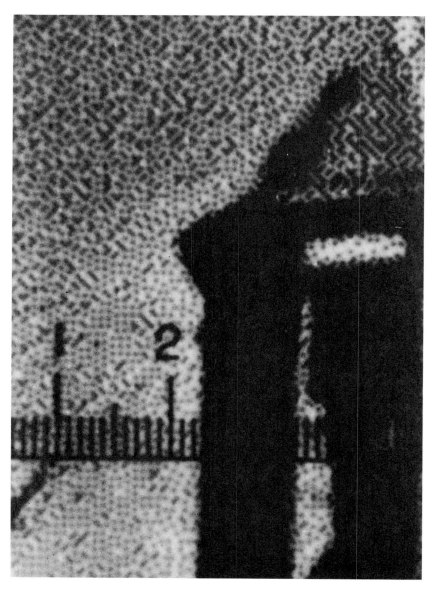

Figure 2.6: Photographic Enlargements of Binary Display Output.
(c) 30× magnification of output from a 285 dots/cm laser recorder.
(Numbered reticle marks are millimeters.)

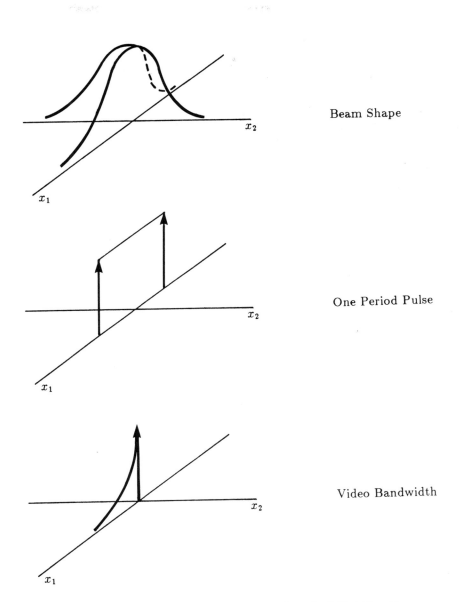

Beam Shape

One Period Pulse

Video Bandwidth

Figure 2.7: Cascaded Components of a "Gaussian" Dot Function.

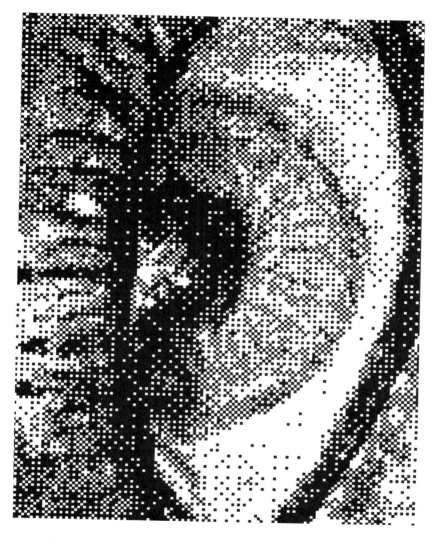

Figure 2.8: Diversity and Nonlinearity in Electrophotographic
Printer Output.
(a) Exact data to be printed.

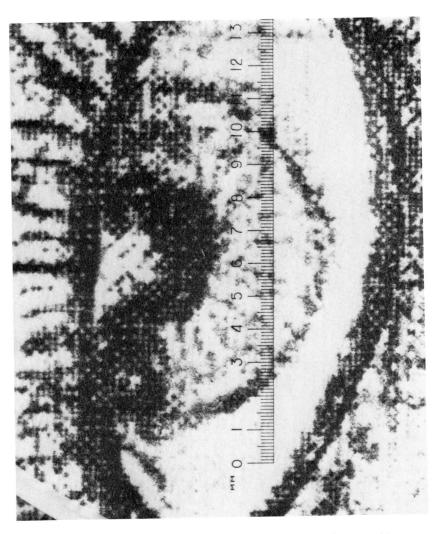

Figure 2.8: Diversity and Nonlinearity in Electrophotographic
Printer Output.
(b) Detail of output from "Product X" (118 dots/cm).
(Numbered reticle marks are millimeters.)

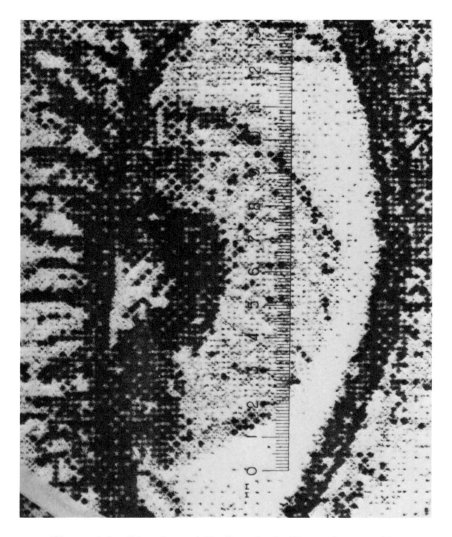

Figure 2.8: Diversity and Nonlinearity in Electrophotographic
Printer Output.
(c) Detail of output from "Product Y" (118 dots/cm).
(Numbered reticle marks are millimeters.)

2.3 Reflectance and Luminance

It is customary to assign zero to the amplitude of a digital image sample
if no output is to be produced at that location. This leads to two conventions; the hardcopy convention assigns zero to the unmarked white
paper, while the convention for luminous displays, such as CRT's, assigns zero to the dark screen.

In this book, the hardcopy convention is adopted, and the gray level,
g, is proportional to reflectance. *Reflectance* is defined as the ratio of
reflected to incident radiant power. $g = 0$ corresponds to the reflectance
of the unmarked white paper, R_W, and $g = 1$ to the reflectance, R_B,
generated by all black output pixels. The desired macroscopic output
reflectance is then

$$R = R_W + g(R_B - R_W) \tag{2.5}$$

For the case of perfectly diffuse reflection copy, the luminance, $\mathcal{L}$,
observed at any angle and given illumination is directly proportional to
the reflectance.

$$\begin{aligned} \mathcal{L} &= \mathcal{L}_W + g(\mathcal{L}_B + \mathcal{L}_W) & (2.6) \\ &= \mathcal{L}_B + (1 - g)(\mathcal{L}_W + \mathcal{L}_B) & (2.7) \end{aligned}$$

For the luminous display convention, equation (2.7) would best describe
the relationship between the desired output luminance, $\mathcal{L}$, and the "gray
level" amplitude, $(1 - g)$.

In either case, if dots do not overlap, then a halftoning scheme should
create the desired output by setting the ratio of black dots to total dots
in a given region of uniform gray level to g. However, dots usually do
overlap.

The fact that the linear relationship of equation (2.5) does not hold
for most devices has been only rarely addressed in the literature. Those
halftoning techniques which have taken dot overlap into account only
consider Class II devices, that is, displays with perfect circular dots
where density does not increase at areas of dot overlap.

Allebach [5] describes a solution for an imaginary device that has
overlapping pixels that are perfectly square and have sizes that are
exact integer multiples of the grid period. Roetling [63] examined one
particular plotter with a fixed dot size. He proposed computing the
overlap into a cell due to all combinations of its eight nearest neighbors,

then using that information to create a compensated classical halftone screen. Similar geometric solutions were proposed to compensate for dot overlap in the error diffusion algorithm, a dispersed-dot technique for rectangular [80] and hexagonal [76] grids.

2.3.1   Direct Measurement

It appears the the most reliable means of compensating for nonlinearity in tone scale (reflectance or luminance) is to use the information from direct measurement of output from a candidate device with a given half-tone technique. The suitability of a clustered of dispersed-dot method can be established, and a compensating tone scale transformation can then be made prior to halftoning.

The reflectance of three hard copy devices was measured and plotted in Figure 2.9, as a function of covering periods of 4×4 addressable locations with increasing number of black output pixels with the dispersed-dot patterns of Figure 6.9(d) (page 147). In each of the three plots, R_W and R_B are slightly different, due to the differences in whiteness of the papers used, and blackness of the maximum densities which can be achieved.

Figure 2.9(a) illustrates the failure of a particular electrophotographic printer to accommodate dispersed-dot halftoning. The inability of this device to reliably print isolated black pixels results in a tone scale which is not only nonlinear but nonmonotonic. Clustered-dot halftoning is necessary for this device.

The devices measured in (b) and (c) of Figure 2.9 can support dispersed-dot halftoning, but are nonlinear. Accounting for dot overlap and even the increase in density at regions of overlap is only part of the reason for this nonlinearity.

To minimize the effect of dot overlap, a 285 dots/cm laser recorder generated the same patterns with "super pixels" composed of 3×3 addressable dots. The output was effectively from a 95 dots/cm device with nearly square dots. The measured reflectance curve for this case is shown on the same plot as the measurements for patterns generated with individual dots in Figure 2.9(c). The disparity from linearity in reflectance was reduced but not nearly to the degree that would be predicted by the reduction in dot overlap.

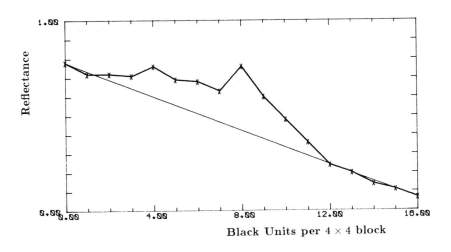

(a) The electrophotographic printer pictured in Figure 2.8(b). It exhibits failure to accommodate dispersed-dot patterns.

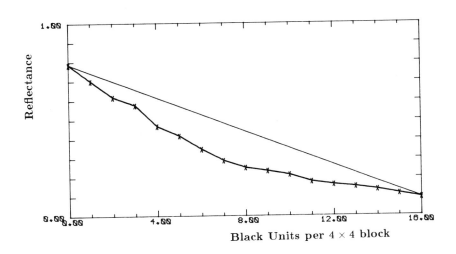

(b) The dot-matrix printer pictured in Figure 2.6(a).

Figure 2.9: Example Reflectance Measurements.

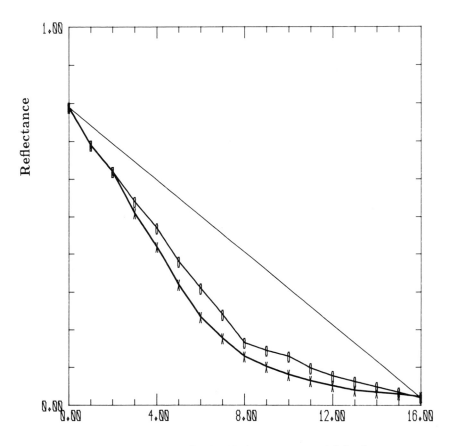

Black Units per 4×4 **block**

"X" for individual dots, "O" for groups of 3×3 dots.

Figure 2.9: Example Reflectance Measurements.
(c) The laser recorder pictured in Figure 2.6(c).
The divergence from linearity is due to more than dot overlap!

If the surface onto which a halftone was printed was completely opaque, then the reflectance could indeed be exactly calculated based on the percentage of area covered. Surprisingly, multiple internal reflections within paper contributes appreciably to the nonlinearity of reflectance, a phenomenon recognized over 30 years ago [15]. This effect depends on the translucency and thickness of the paper, as well as the size and distribution of dots on the top surface.

Considering the complexity and number of parameters that contribute to the relationship between gray level, g, and output reflectance, R, the best method of calibrating tone scale is direct measurement.

Chapter 3

Tools for Fourier Analysis

The Fourier transform has been employed in the past to evaluate half-toned images, but only very special cases of even period ordered dither on square grids were considered [3,37,59]. In this chapter, comprehensive methods for analyzing the nature of all types of patterns produced by halftoning will be developed in the frequency domain for both rectangular and hexagonal grids.

The most characteristic feature of a halftone technique is the texture generated in areas of uniform gray. The rendition of high frequency detail depends primarily on how sharp the image was (or to what extent high pass filtering was performed) prior to halftoning. As stated earlier, some degree of presharpening will usually produce higher quality halftoned pictures.

The best measure of the virtues of a halftone algorithm, then, is its ability to render areas of uniform gray. The approach used to examine this ability in the frequency domain depends on whether or not the resulting binary texture patterns are periodic. This chapter is divided into two part to address each case separately. Introduced are "exposure plots" of *composite Fourier transforms* which will present insight into the nature of the periodic output of ordered dither, and *radially averaged power spectra* along with a measure of anisotropy to provide a mechanism for studying aperiodic patterns.

3.1 Periodic (Ordered) Patterns

The ordered dither algorithms of Chapters 5, 6, and 7 halftone by thresholding or "screening" with periodic threshold arrays. The binary output from such halftone processes will also be periodic with the same spatial period as the threshold array. The spatial period will be specified by two vectors, $\mathbf{p}_1$ and $\mathbf{p}_2$, in terms of the spatial sampling vectors, $\mathbf{v}_1$ and $\mathbf{v}_2$, described in section 2.1.1.

The spatial periods can be thought of as tiles which cover all of two-space. Two types of periods are of interest. Figure 3.1 shows examples of *odd* and *even periods*[1] (tiles) on semiregular rectangular and hexagonal grids of the first kind. Since rectangular tiles share each vertex with 4 other tiles, and each edge with 2 other tiles, only 1 vertex and 2 edges are unique to each tile. By a similar argument, only 2 vertices and 3 edges are unique to each hexagonal tile. The outlined edge and vertices in Figure 3.1 show those points which are not part of the unique period.

Even periods are replicated by period vectors that are collinear with the sampling vectors,

$$\begin{aligned}
\mathbf{p}_1 &= N\mathbf{v}_1 \\
\mathbf{p}_2 &= N\mathbf{v}_2,
\end{aligned} \tag{3.1}$$

for some integer, N. Odd periods are defined by

$$\begin{aligned}
\mathbf{p}_1 &= N(\mathbf{v}_1 + \mathbf{v}_2) \\
\mathbf{p}_2 &= N(\mathbf{v}_2 - \mathbf{v}_2).
\end{aligned} \tag{3.2}$$

It is important to note that two odd periods on a rectangular grid and three odd periods on a hexagonal grid can always be packed into one even period.

The derivation of a general expression for the Fourier transform of periodic patterns with an even period is addressed in the following section.

---

[1]Describing periods as "odd" or "even" is consistent with period *order*, η, introduced in Chapter 6.

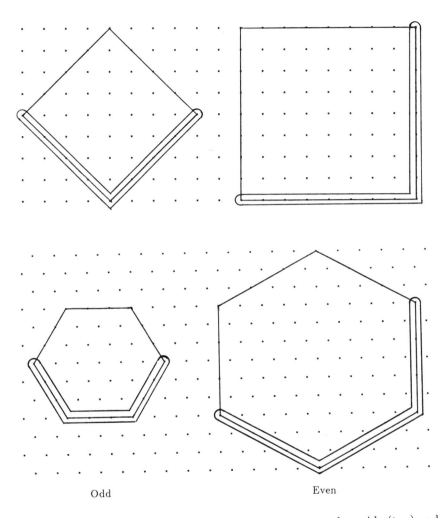

Odd   Even

Figure 3.1: Odd and Even Spatial Periods for rectangular grids (top) and hexagonal grids of the first kind (bottom). Boundary locations which are not part of the unique period are outlined.

3.1.1   Continuous-space Fourier Transform Computation

For nonrectangular grids, it is not immediately clear how a discrete Fourier transform can be computed, or what its dimensions in continuous frequency space are. So, a method of computation along with an explicit expression in continuous space is sought.

Probably the first formalization of the Fourier representation of nonrectangularly sampled spaces was by Petersen and Middleton in 1962 [56]. Mersereau later specifically addressed hexagonally sampled signals with his derivation of the Hexagonal Discrete Fourier Transform (HDFT) [50]. But this expression is complicated, primarily because hexagonal periods can not in general be rearranged to repeat in a rectangular (rhomboidal) fashion.

For the case where the number of elements in a hexagonal period, or any shaped period, on a general periodic sampling grid is a perfect square, the following is a proof inspired by the multidimensional sampling theorem [18] that the canonical rectangular DFT can be used to compute its Fourier transform.

Proof

The two-dimensional continuous space Fourier transform, $C(f_1, f_2)$, of an image, $c(x_1, x_2)$, can be expressed as

$$\begin{aligned}
C(f_1, f_2) &= \mathcal{F}\{c(x_1, x_2)\} \\
&= \int_{-\infty}^{\infty} \int_{-\infty}^{\infty} c(x_1, x_2) e^{-j2\pi(f_1 x_1 + f_2 x_2)} dx_1 dx_2 \\
c(x_1, x_2) &= \mathcal{F}^{-1}\{C(f_1, f_2)\} \\
&= \int_{-\infty}^{\infty} \int_{-\infty}^{\infty} C(f_1, f_2) e^{j2\pi(f_1 x_1 + f_2 x_2)} df_1 df_2
\end{aligned}$$

or, more conveniently in vector form:

$$C(\mathbf{f}) = \mathcal{F}\{c(\mathbf{x})\} = \int_{-\infty}^{\infty} c(\mathbf{x}) e^{-j2\pi \mathbf{f}^\top \mathbf{x}} d\mathbf{x} \qquad (3.3)$$

$$c(\mathbf{x}) = \mathcal{F}^{-1}\{C(\mathbf{f})\} = \int_{-\infty}^{\infty} C(\mathbf{f}) e^{j2\pi \mathbf{f}^\top \mathbf{x}} d\mathbf{f} \qquad (3.4)$$

The units of the frequency components in these expressions are cycles/unit-length as opposed to the more common radians/unit-length.

One important identity that will be used is the transform of a two-dimensional impulse train:

$$\mathcal{F}\left\{\sum_{\mathbf{n}}^{\infty}\delta(\mathbf{x}-\mathbf{An})\right\} = \frac{1}{|\det \mathbf{A}|}\sum_{\mathbf{m}}^{\infty}\delta(\mathbf{f}-\mathbf{Bm}) \qquad (3.5)$$

where
$$\mathbf{B}^{\mathsf{T}}\mathbf{A} = \mathbf{I},$$
$$\det \mathbf{A} = a_{11}a_{22} - a_{12}a_{21},$$

and
$$\sum_{\mathbf{n}}^{\infty} \text{ represents } \sum_{n_1=-\infty}^{\infty}\sum_{n_2=-\infty}^{\infty}.$$

Note that the spatial pixel area as defined in section 2.1.1 is equal to $|\det \mathbf{A}|$, the frequency "pixel area" is $|\det \mathbf{B}|$, and $|\det \mathbf{A}| = |\det \mathbf{B}|^{-1}$. The matrix $\mathbf{B}$ can be expressed explicitly as

$$\mathbf{B} = (\mathbf{A}^{-1})^{\mathsf{T}} = \frac{1}{\det \mathbf{A}}\begin{bmatrix} a_{22} & -a_{21} \\ -a_{12} & a_{11} \end{bmatrix}. \qquad (3.6)$$

Recalling that $\mathbf{V}$ has been defined as the *spatial sampling matrix* (equations (2.2) and (2.3), page 18),

$$\mathcal{F}\left\{\sum_{\mathbf{n}}^{\infty}\delta(\mathbf{x}-\mathbf{Vn})\right\} = \frac{1}{|\det \mathbf{V}|}\sum_{\mathbf{m}}^{\infty}\delta(\mathbf{f}-\mathbf{Um}) \qquad (3.7)$$

where $\mathbf{U} = [\mathbf{u}_1 \vdots \mathbf{u}_2] = (\mathbf{V}^{-1})^{\mathsf{T}}$ is the *frequency baseband replication matrix*.

Suppose $c(\mathbf{x})$ is a periodic array of weighted impulses on the sampling grid of equation (2.2) with an $N \times N$ rhomboidal shaped period. It could be expressed as

$$c(\mathbf{x}) = \left(\sum_{\mathbf{n}}^{N} I[\mathbf{n}]\delta(\mathbf{x}-\mathbf{Vn})\right) * \sum_{\mathbf{l}}^{\infty}\delta(\mathbf{x}-\mathbf{Pl}) \qquad (3.8)$$

where "$*$" denotes two-dimensional convolution,

$$\sum_{\mathbf{n}}^{N} \text{ represents } \sum_{n_1=0}^{N}\sum_{n_2=0}^{N},$$

$I[\mathbf{n}]$ is a discrete-space image, and $\mathbf{P} = [\mathbf{p}_1 \vdots \mathbf{p}_2] = N\mathbf{V}$ is the *spatial period replication matrix*.

Note that

$$\mathcal{F}\left\{ \sum_{\mathbf{l}}^{\infty} \delta(\mathbf{x} - \mathbf{Pl}) \right\} = \frac{1}{|\det \mathbf{P}|} \sum_{\mathbf{k}}^{\infty} \delta(\mathbf{f} - \mathbf{Qk}) \qquad (3.9)$$

where $\mathbf{Q} = [\mathbf{q}_1 \vdots \mathbf{q}_2] = (\mathbf{P}^{-1})^{\top}$ is the *frequency sampling matrix*. $\mathbf{Q}$ also equals $\frac{1}{N}\mathbf{U}$, since $\mathbf{P} = N\mathbf{V}$ and $\mathbf{U} = (\mathbf{V}^{-1})^{\top}$.

The Fourier transform of this image is

$$
\begin{aligned}
C(\mathbf{f}) \ &= \ \mathcal{F}\{c(\mathbf{x})\} \\
&= \ \left(\int_{-\infty}^{\infty} \sum_{\mathbf{n}}^{N} I[\mathbf{n}]\delta(\mathbf{x} - \mathbf{Vn}) \, e^{-j2\pi \mathbf{f}^{\top}\mathbf{x}} dx \right) \qquad (3.10) \\
&\qquad \times \frac{1}{|\det \mathbf{P}|} \sum_{\mathbf{k}}^{\infty} \delta(\mathbf{f} - \mathbf{Qk}) \\
&= \ \left(\sum_{\mathbf{n}}^{N} I[\mathbf{n}]e^{-j2\pi \mathbf{f}^{\top}\mathbf{Vn}} \right) \frac{1}{|\det \mathbf{P}|} \sum_{\mathbf{k}}^{\infty} \delta(\mathbf{f} - \mathbf{Qk}). \ (3.11)
\end{aligned}
$$

This expression is zero everywhere except at $\mathbf{f} = \mathbf{Qk}$, so the frequency term in the exponent becomes

$$\mathbf{f}^{\top} = \mathbf{k}^{\top}\mathbf{Q}^{\top}$$

where $\mathbf{Q}^{\top} = \frac{1}{N}\mathbf{V}^{-1}$ since $\mathbf{Q} = (\mathbf{P}^{-1})^{\top}$ and $\mathbf{P} = N\mathbf{V}$.

Thus, equation (3.11) becomes simply

$$C(\mathbf{f}) \ = \ \frac{1}{|\det \mathbf{P}|} \sum_{\mathbf{k}}^{\infty} \left(\sum_{\mathbf{n}}^{N} I[\mathbf{n}]e^{-j\frac{2\pi}{N}\mathbf{k}^{\top}\mathbf{n}} \right) \delta(\mathbf{f} - \mathbf{Qk}) \qquad (3.12)$$

or,

$$C(\mathbf{f}) \ = \ \frac{1}{|\det \mathbf{P}|} \sum_{\mathbf{k}}^{\infty} I[\mathbf{k}]\delta(\mathbf{f} - \mathbf{Qk})$$

where, in scalar notation,

$$I[\mathbf{k}] \ = \ I[k_1, k_2] \ = \ \sum_{n_1=0}^{N} \sum_{n_2=0}^{N} I[n_1, n_2]e^{-j\frac{2\pi}{N}(k_1 n_1 + k_2 n_2)} \qquad (3.13)$$

is recognized as the familiar two-dimensional discrete Fourier transform! Equation (3.12) satisfies the need for an explicit expression in continuous frequency space and a simple mechanism for computation.

Check of Proof

For completeness and as a check for this proof, the reverse Fourier transform can be computed in a similar way. Since $I[\mathbf{k}]$ is also periodic with period $N \times N$,

$$C(\mathbf{f}) \;=\; \frac{1}{|\det \mathbf{P}|} \left(\sum_{\mathbf{k}}^{N} I[\mathbf{k}]\delta(\mathbf{f} - \mathbf{Q}\mathbf{k}) \right) \;*\; \sum_{\mathbf{m}}^{\infty} \delta(\mathbf{f} - \mathbf{U}\mathbf{m}) \qquad (3.14)$$

and its inverse Fourier transform is

$$
\begin{aligned}
c(\mathbf{x}) \;&=\; \mathcal{F}^{-1}\{C(\mathbf{f})\} \\
&=\; \frac{1}{|\det \mathbf{P}|} \left(\int_{-\infty}^{\infty} \sum_{\mathbf{k}}^{N} I[\mathbf{k}]\delta(\mathbf{f} - \mathbf{Q}\mathbf{k})\, e^{j2\pi \mathbf{f}^{\top}\mathbf{x}} df \right) \qquad (3.15) \\
&\quad \times |\det \mathbf{V}| \sum_{\mathbf{n}}^{\infty} \delta(\mathbf{x} - \mathbf{V}\mathbf{n}) \\
&=\; \frac{|\det \mathbf{V}|}{|\det \mathbf{P}|} \sum_{\mathbf{n}}^{\infty} \left(\sum_{\mathbf{k}}^{N} I[\mathbf{k}] e^{j2\pi \mathbf{k}^{\top}\mathbf{n}} \right) \delta(\mathbf{x} - \mathbf{V}\mathbf{n}) \qquad (3.16)
\end{aligned}
$$

where $\mathbf{Q}^{\top} = \frac{1}{N}\mathbf{V}^{-1}$ again simplifies the exponent.

The expression is further simplified by observing that $|\det \mathbf{P}| = N^2 |\det \mathbf{V}|$, so

$$c(\mathbf{x}) = \sum_{\mathbf{n}}^{\infty} I[\mathbf{n}]\delta(\mathbf{x} - \mathbf{V}\mathbf{n}) = \left(\sum_{\mathbf{n}}^{N} I[\mathbf{n}]\delta(\mathbf{x} - \mathbf{V}\mathbf{n}) \right) \;*\; \sum_{\mathbf{l}}^{\infty} \delta(\mathbf{x} - \mathbf{P}\mathbf{l})$$
$$(3.17)$$

because $I[\mathbf{n}]$ is periodic with period $N \times N$, where in scalar notation

$$I[\mathbf{n}] \;=\; I[n_1, n_2] \;=\; \frac{1}{N^2} \sum_{k_1=0}^{N} \sum_{k_2=0}^{N} I[k_1, k_2] e^{j\frac{2\pi}{N}(k_1 n_1 + k_2 n_2)} \qquad (3.18)$$

is the two-dimensional inverse discrete Fourier transform. $\qquad \square$

This proof applies to periodic grids in general. For the case of semi-iregular grids considered in this text, the matrices involved can be expressed in terms of the sample and line periods.

For the rectangular case:

$$\mathbf{V} = \begin{bmatrix} S_r & 0 \\ 0 & L_r \end{bmatrix} \quad \mathbf{U} = \begin{bmatrix} (S_r)^{-1} & 0 \\ 0 & (L_r)^{-1} \end{bmatrix} \tag{3.19}$$

$$\mathbf{P} = \begin{bmatrix} NS_r & 0 \\ 0 & NL_r \end{bmatrix} \quad \mathbf{Q} = \begin{bmatrix} (NS_r)^{-1} & 0 \\ 0 & (NL_r)^{-1} \end{bmatrix} \tag{3.20}$$

For the semiregular hexagonal case:

$$\mathbf{V} = \begin{bmatrix} S_h & S_h/2 \\ 0 & L_h \end{bmatrix} \quad \mathbf{U} = \begin{bmatrix} (S_h)^{-1} & 0 \\ (-2L_h)^{-1} & (L_h)^{-1} \end{bmatrix} \tag{3.21}$$

$$\mathbf{P} = \begin{bmatrix} NS_h & NS_h/2 \\ 0 & NL_h \end{bmatrix} \quad \mathbf{Q} = \begin{bmatrix} (NS_h)^{-1} & 0 \\ (-2NL_h)^{-1} & (NL_h)^{-1} \end{bmatrix} \tag{3.22}$$

The associated vectors are displayed in figures 3.2 and 3.3. Recalling that equation (3.12) is only valid for periods which have a perfect square number of elements, only even tiles meet that condition. This is most certainly true for rectangular tiles as shown in Figure 3.2, but Figure 3.3 illustrates why the condition can be met for hexagonal grids. The rhomboidal tiles shown cover the plane with precisely the same data as the hexagonal tiles.

In frequency space, the rhombus described by the vectors, $\mathbf{u}_1$ and $\mathbf{u}_2$, define how the Fourier transform is tiled but does not necessarily describe the shape of the baseband. In the hexagonal case (Figure 3.3), the baseband has the hexagonal shape as shown. It is interesting to note that when the period is even, a spatial hexagonal grid of the first kind has a transform on a frequency hexagonal grid of the second kind (and vice versa).

For the case where the period is odd, equation (3.12) can still be used by invoking the Similarity Theorem [12, p. 370]. Figure 3.4 shows how 2 odd rectangular tiles and 3 odd hexagonal tiles can be packed into a single even period. The Similarity Theorem states that when exactly K periods are transformed as one period, each nonzero frequency component in the resulting DFT will be accompanied by K zero coefficients and have a magnitude K times as large as the DFT of a single period.

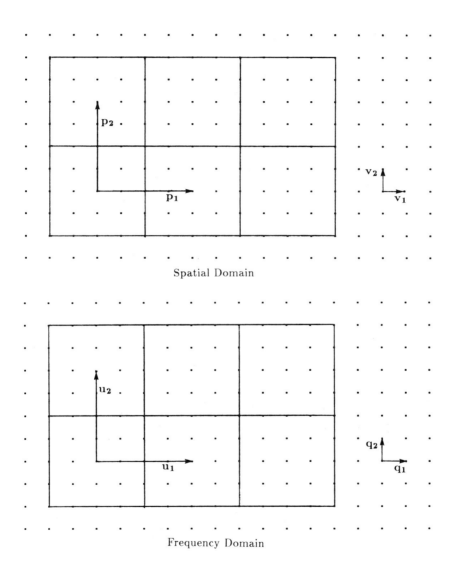

Spatial Domain

Frequency Domain

Figure 3.2: Rhomboidal Tiling of a Rectangular Array
coincides with the rectangularly shaped period when the period is even.

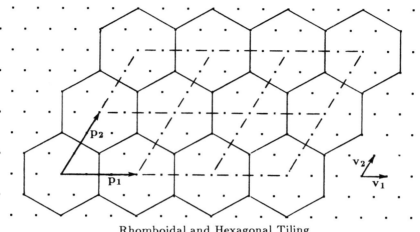

Rhomboidal and Hexagonal Tiling
of equivalent data in the Spatial Domain.

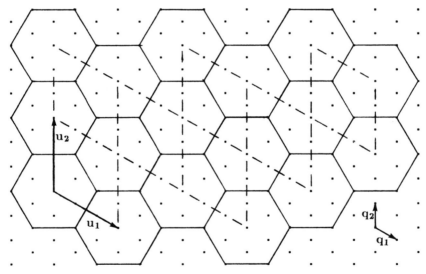

Resulting Rhomboidal tiling in the Frequency Domain.
Baseband has Hexagonal Shape.

Figure 3.3: Rhomboidal Tiling of a Hexagonal Array with an even period.

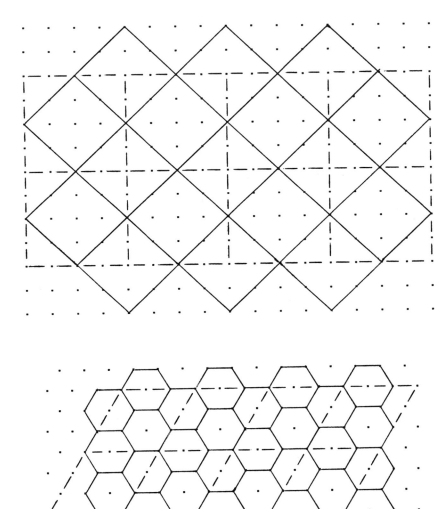

Figure 3.4: Packing an integral number of odd periods into an even period.

So, packing odd tiles in this way only requires the additional step of dividing the resulting DFT by K.

This step is automatically handled by the normalization term in equation (3.12), $|\det \mathbf{P}|^{-1}$, where

$$|\det \mathbf{P}| = Z \times (\text{pixel area}) = (\text{spatial period area}).$$

where Z is the number of elements in the period used.

3.1.2 Composite Fourier Transform

The DFT is a strictly valid description of the Fourier transform *only* for periodic arrays of weighted impulses of which equation (3.8) is one. It is impossible for a signal to be both of finite spatial extent *and* truly band limited. When the DFT is applied to real signals, some degree of spectral overlap must be accepted. In well designed systems, this error is made inappreciable, but never zero.

In this section, the Fourier transform of the images resulting from halftoning a two-dimensional plane of one gray level are examined. $I[\mathbf{n}]$ in this case happens to consist of only 1's and 0's. Such images are precisely as described by equation (3.8), and thus the results are theoretically exact. (This could be the only practical use of the DFT that can make that claim.)

The output binary image is capable of rendering one of $Z + 1$ gray levels for ordered dither with a Z element threshold array. The binary image, $I[\mathbf{n}; g]$, resulting from halftoning a continuous-tone image consisting of one constant gray level, g, with a given threshold array has a DFT equation (3.13) denoted by $I[\mathbf{k}; g]$.

The phase of $I[\mathbf{n}]$ relative to an origin is not important. Of interest is the magnitude of $I[\mathbf{k}; g]$, which will provide insights into the relative distribution of energy in frequency space. To summarize this information over all gray levels for a given threshold matrix, an average is used. The *composite DFT* is defined as having frequency components,

$$I_\Sigma[\mathbf{k}] \equiv \frac{1}{Z+1} \sum_{\text{all } g} |I[\mathbf{k}; g]| \tag{3.23}$$

Plugging $I_\Sigma[\mathbf{k}]$ into equation (3.12) yields a specification of the location and magnitude of Fourier transform impulses. This *composite*

Fourier transform is defined as

$$C_\Sigma(\mathbf{f}) = \frac{1}{|\det \mathbf{P}|} \sum_{\mathbf{k}}^{\infty} I_\Sigma[\mathbf{k}]\delta(\mathbf{f} - \mathbf{Q}\mathbf{k}) \qquad (3.24)$$

Note that in all cases, the zero frequency term of $C_\Sigma(\mathbf{f})$ will be an impulse with area

$$\frac{I_\Sigma[\mathbf{0}]}{|\det \mathbf{P}|} = \frac{1}{2SL}, \qquad (3.25)$$

since $I_\Sigma[\mathbf{0}]$ will always equal $Z/2$, the average number of black output pixels in a period.

A reasonable means of displaying $C_\Sigma(\mathbf{f})$, is with dots of an area proportional to the magnitude of the impulses. Such a display is similar to a photograph of an ideal optical Fourier transform consisting of points of light of different intensity; the resulting size of the exposed points would be proportional to their magnitude. Thus, the name "exposure plots" will be used to describe them. An exposure plot of the composite Fourier transform will be presented with each ordered dither threshold array explored in Chapters 5, 6, and 7.

3.2 Aperiodic Patterns

Halftone processes which do not produce output by thresholding with a deterministic, periodic threshold array will in general be aperiodic. The analysis of section 3.1 will be inappropriate. Such aperiodic dither patterns can be modeled as stochastic processes.

The unconditional probability mass function of any individual binary output pixel, $I[\mathbf{n}]$, is

$$p_I(I[\mathbf{n}]) = \begin{cases} g & \text{for} \quad I[\mathbf{n}] = 1 \\ (1-g) & \text{for} \quad I[\mathbf{n}] = 0. \end{cases} \qquad (3.26)$$

Since this is true for all $\mathbf{n}$, $I[\mathbf{n}]$ is a stationary random process with

$$\mathrm{E}\{I[\mathbf{n}]\} = g \qquad (3.27)$$

$$\text{and,} \qquad \mathrm{var}\{I[\mathbf{n}]\} \equiv \sigma_g^2 = g(1-g). \qquad (3.28)$$

The mean of $I[\mathbf{n}]$ is exactly what is expected, since the gray level g is being represented. The variance of $I[\mathbf{n}]$ varies with g, and has a maximum at $g = \frac{1}{2}$, midway between the extremes of zero variance at solid black and white.

3.2.1 Estimating the Power Spectrum

The Fourier transform of the autocorrelation function of a stationary random process is the Power Spectrum, $P(\mathbf{f})$. In most cases, the autocorrelation function of a given aperiodic halftone process will not be known, so a method of spectral estimation must be employed to produce an estimate, $\hat{P}(\mathbf{f})$, of $P(\mathbf{f})$. Bartlett's Method [9] of averaging periodograms, named after the one who first suggested the technique for the one-dimensional case, will be used in this study to produce $\hat{P}(\mathbf{f})$.

A periodogram is the magnitude squared of the Fourier transform of sample output, $I[\mathbf{n}; g]$, divided by the sample size. All spectral estimates in this text will be produced by averaging 10 periodograms of 256×256 output pixels from a given halftone rendering of a fixed gray level. Figure 3.5 illustrates how the sample output will be segmented for the purpose of computing the 10 periodograms for both rectangular and hexagonal grids. Since some of the processes of Chapter 8 have transient behavior near edges or boundaries, the segments are cropped sufficiently far from output edges to avoid such artifacts; only the "steady state" output will be measured.

It can be shown [53] that a spectral estimate formed by averaging K periodograms has an expectation equal to $P(\mathbf{f})$ smoothed by convolution with the Fourier transform of a triangle function with a span equal to the size of the sample segments, and variance

$$\text{var}\{\hat{P}(\mathbf{f})\} \approx \frac{1}{K} P^2(\mathbf{f}). \tag{3.29}$$

3.2.2 Radially Averaged Power Spectra and Anisotropy

A desirable attribute of a well produced aperiodic halftone of a fixed gray level is radial symmetry; directional artifacts are perceptually disturbing. $\hat{P}(\mathbf{f})$ is a function of two dimensions. Although anisotropies

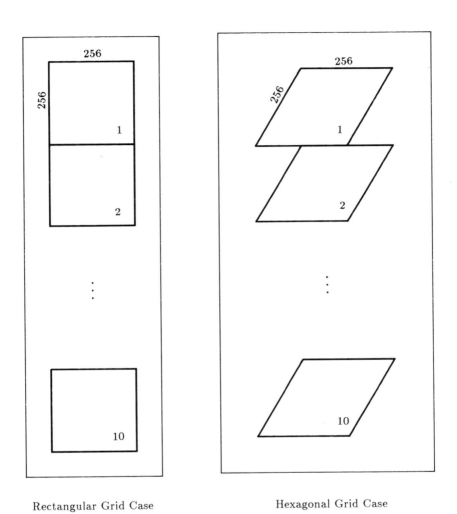

Rectangular Grid Case   Hexagonal Grid Case

Figure 3.5: Segmentation Strategy for Spectral Estimation.

in $I[\mathbf{n}; g]$ can be qualitatively observed by studying 3-D plots of $\hat{P}(\mathbf{f})$, a more quantitative metric is proposed.

Figure 3.6 shows how spectral estimates, $\hat{P}(\mathbf{f})$, can be partitioned into annuli of width Δ for regular (a) rectangular and (b) hexagonal grids. Each annulus has a central radius f_r, the radial frequency, and $N_r(f_r)$ frequency samples.

Two useful one-dimensional statistics can be derived from averages within these annuli. The sample mean of the frequency samples of $\hat{P}(\mathbf{f})$ in the annulus, $||\mathbf{f}| - f_r| < \Delta/2$ about f_r, is defined as the *radially averaged power spectrum*,

$$P_r(f_r) = \frac{1}{N_r(f_r)} \sum_{i=1}^{N_r(f_r)} \hat{P}(\mathbf{f}). \tag{3.30}$$

The sample variance of the same frequency samples is defined as

$$s^2(f_r) = \frac{1}{N_r(f_r) - 1} \sum_{i=1}^{N_r(f_r)} (\hat{P}(\mathbf{f}) - P_r(f_r))^2. \tag{3.31}$$

Note that the sum is divided by $N_r(f_r) - 1$ and not $N_r(f_r)$, so as to yield an unbiased estimate of the variance (see [22] or [54]).

For each gray level output of a halftone process to be analyzed, two plots will be presented. First, the radially averaged power spectrum divided by σ_g^2 will be shown. Since spectral energy increases with σ_g^2 equation (3.28), normalizing $P_r(f_r)$ by this amount will render all plots on the same relative scale. Because of the importance of σ_g^2, its relationship to gray level, g, is now shown in Figure 3.7.

Secondly, the anisotropy of $\hat{P}(\mathbf{f})$ will be plotted. *Anisotropy* is defined in this book as

$$\frac{s^2(f_r)}{P_r^2(f_r)}, \tag{3.32}$$

a measure of the relative variance of frequency samples within a given annulus. Note that this measure can be described as the square of the *coefficient of variation*, or as a "noise-to-signal" ratio. Because the range of anisotropy values can be quite large, it will be plotted in decibels.

The zero frequency term is assumed to be close to g in all cases; the spike at this frequency will not be shown since it does not contribute to the structure of the dither pattern.

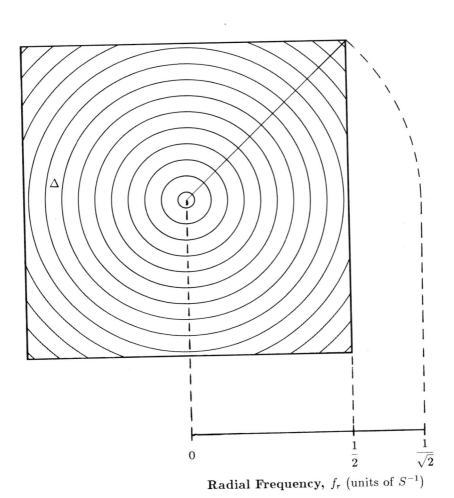

Radial Frequency, f_r (units of S^{-1})

Figure 3.6: Segmenting the Spectral Estimate into Concentric Annuli.
(a) Regular Rectangular Grid, $S = L$ ($\alpha = 1$).
181 annuli actually used.

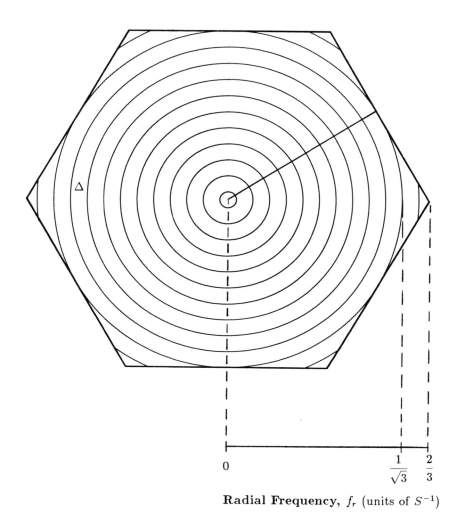

Radial Frequency, f_r (units of S^{-1})

Figure 3.6: Segmenting the Spectral Estimate into Concentric Annuli.
(b) Regular Hexagonal Grid, $S = \frac{2}{\sqrt{3}}L$ ($\alpha = \frac{2}{\sqrt{3}}$).
148 annuli actually used.

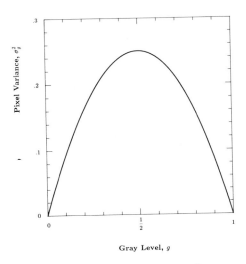

Figure 3.7: The dependence of σ_g^2 on gray level.

3.2.2.1 Quality of Measurement

To what extent will $P_r(f_r)$ and $s^2(f_r)/P_r^2(f_r)$ be meaningful metrics? From equation (3.29) and the fact that $K = 10$ segments are used in the estimate, $\hat{P}(\mathbf{f})$,

$$\frac{\text{var}\{\hat{P}(\mathbf{f})\}}{P^2(\mathbf{f})} \approx \frac{1}{10}. \tag{3.33}$$

If $P(\mathbf{f})$ is perfectly radially symmetric, the measure of anisotropy, $s^2(f_r)/P_r^2(f_r)$, is merely an estimate of the above ratio. Therefore, an anisotropy of $\frac{1}{10}$ or -10 dB should be considered "background noise", and a reference line at this level will appear in each plot.

Also, if anisotropy is low, that is, close to -10 dB, indicating good radial symmetry, then $\hat{P}(\mathbf{f})$ is effectively a function of one independent variable, f_r, instead of two variables, $\mathbf{f}$. The variance of $P_r(f_r)$ is that of equation (3.33) divided by $N_r(f_r)$, assuming that each of the $N_r(f_r)$ samples are independent. This reduction in variance as $N_r(f_r)$ increases is indeed observed in the experimental data of Chapters 4 and 8.

$N_r(f_r)$ depends on the width of the annuli, Δ. As indicated earlier, in this book all estimates, $\hat{P}(\mathbf{f})$, will consist of 256^2 frequency samples. The size of Δ was chosen so that exactly one sample along each frequency axis fell into each annulus; that is, $\Delta = |\mathbf{q}|$, where $\mathbf{q} = \mathbf{q}_1 = \mathbf{q}_2$, since the grids are assumed to be regular. A plot of $N_r(f_r)$ for (a) rectangular and (b) hexagonal grids is plotted in Figure 3.8.

The irregularities in the shape of these plots are a consequence of rectangular and hexagonal grids not being perfectly radially symmetric. The number of grid points that fall into a particular annulus essentially increases linearly, as one would expect, up to the largest annulus that will completely fit within the shape of the baseband; this occurs at $f_r/S^{-1} = \frac{1}{2}$ for rectangular grids and $f_r/S^{-1} = \frac{1}{\sqrt{3}}$ for hexagonal grids.

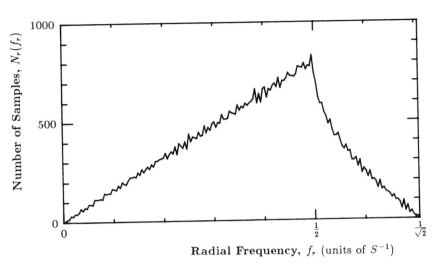

(a) Regular Rectangular Grid.

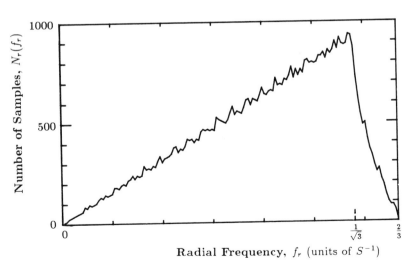

(b) Regular Hexagonal Grid.

Figure 3.8: Number of Frequency Samples within each Annulus.

Chapter 4

Dithering with White Noise

In this chapter, the process of creating a dispersed-dot halftone by the point process of thresholding an input image with uniformly distributed, uncorrelated (white) noise is investigated. The quality of output from this method does not deserve consideration for practical use; as will be seen in Chapter 6, another point process taking no more computational effort performs much better than this one.

So why should a chapter be devoted to this so called technique of "random dither"? There are two reasons.

The first is historical. The idea was the first used to exploit the fact that electronic displays can have independently addressable dots. Goodall [26] in 1951 and Roberts [58] in 1962 demonstrated how contouring due to insufficient gray levels can be corrected by adding noise of this type. This is perhaps the first technique that comes to mind to correct the shortcomings of using a fixed threshold, and in the early days of digital halftoning it was always referenced for comparison; in fact the name "ordered dither" was meant to contrast random dither.

Secondly, investigating the output of single gray levels, $I[\mathbf{n}; g]$, dithered in this way, provides a means to check the validity of the newly introduced metrics of radially averaged power spectrum and anisotropy. Since $I[\mathbf{n}; g]$ is white noise, it has a known autocorrelation function, namely an impulse at the origin with area σ_g^2. So, the power spectrum should be radially symmetric with fixed amplitude, σ_g^2. Such radial symmetry has been observed optically in Fraunhofer diffraction patterns of randomly distributed apertures [29].

The random numbers used in this chapter (and in Chapter 8) are, strictly speaking, pseudorandom. They are produced by means of a *multiplicative congruential random number generator* [39] available with many programming libraries. This is a very efficient scheme requiring only one multiplication and division per random number, and in the case used in this study, has a repeat period of 2^{32}.

The effects of dithering with white noise on regular rectangular and hexagonal grids will now be considered separately.

4.1   Rectangular Grids

A random dithered gray scale ramp is shown in Figure 4.1. Examples of the effect of random dither on a scanned and synthesized image are given in Figures 4.2 and 4.3. They suffer from a grainy appearance. This is the case at any displayed resolution because of the presence of long wavelengths (low frequencies) at all gray levels.

The radially averaged power spectrum and anisotropy for gray levels, (a) $g = \frac{1}{8}$, (b) $g = \frac{1}{2}$, (c) $g = \frac{7}{8}$ are displayed in Figure 4.4. With each plot a small portion of the sample image, $I[\mathbf{n}; g]$, is shown at the top. The well behaved nature of these plots validate four things:

1. The amplitude of $P_r(f_r)$ is correct, that is, flat, as expected for white noise.

2. The values of σ_g^2 are as predicted by equation (3.28), Figure 3.7 (page 59).

3. The apparent variance of $P_r(f_r)$ decreases with $N_r(f_r)$ (Figure 3.8) with a minimum at $\frac{1}{2}$, ($\frac{1}{\sqrt{3}}$ for the hexagonal case).

4. The anisotropy measure is correct. White noise is radially symmetric so the anisotropy should be at the "background noise" minimum of -10 dB.

The characteristic features seen here will serve as a reference for the many plots to be studied in Chapter 8.

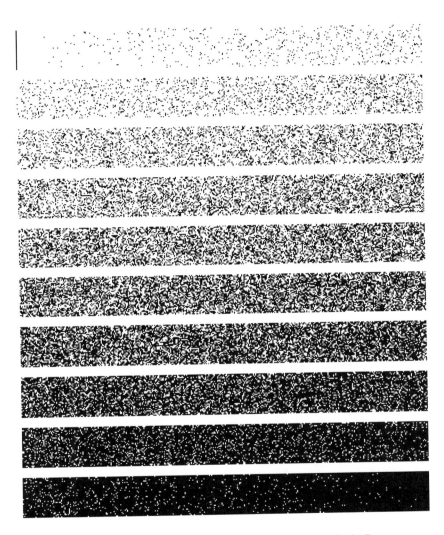

Figure 4.1: Rectangular Random Dither of a Gray Scale Ramp.

Figure 4.2: Rectangular Random Dither of a Scanned Picture.

Figure 4.3: Rectangular Random Dither of a Synthesized Image.

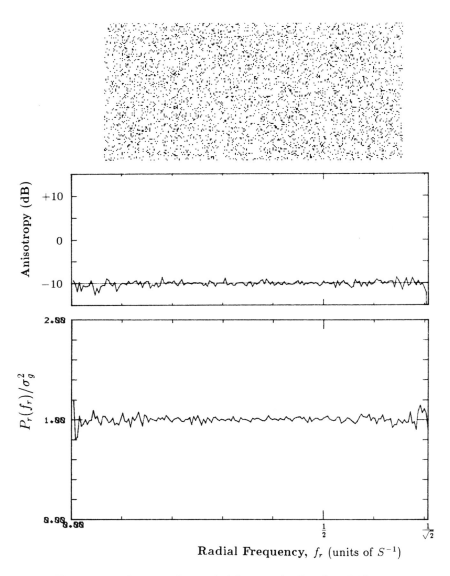

Figure 4.4: Rectangular Radial Spectra for Random Dither.
(a) $\boxed{g = \frac{1}{8}}$, $\sigma_g^2 = \frac{7}{64} \approx .1094$

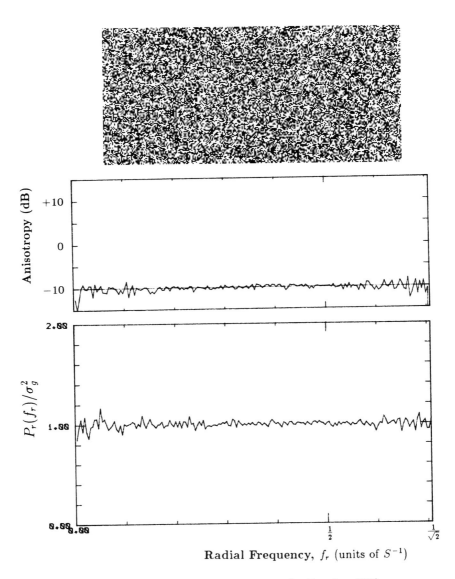

Figure 4.4: Rectangular Radial Spectra for Random Dither.

(b) $\boxed{g = \frac{1}{2}}$, $\sigma_g^2 = \frac{1}{4} = .25$

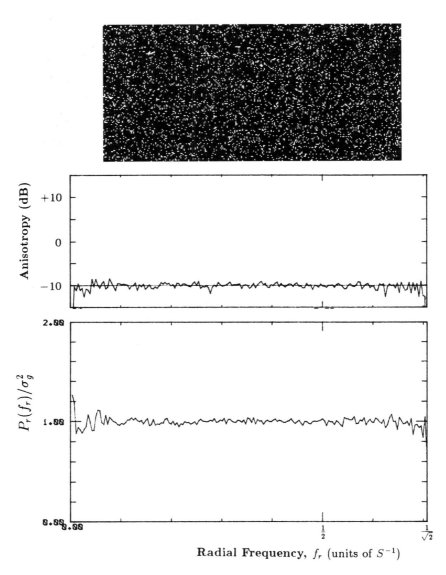

Figure 4.4:   Rectangular Radial Spectra for Random Dither.
(c) $\boxed{g = \frac{7}{8}}$, $\sigma_g^2 = \frac{7}{64} \approx .1094$

4.1.1 The Mezzotint

It should be mentioned that a randomly dithered binary image is often called a *mezzotint*, after a print making technique invented in the seventeenth century [33]. The etymology of this Italian-rooted word is that of "halftone", a word used to differentiate the modern mechanical printing technique from the older art. The dark regions of an image were roughened or ground on a copper plate by a skilled craftsman in a somewhat random fashion by hand. The resulting scratches acted as tiny wells which held ink, much the same as in modern day gravure.

A photographic enlargement detailing an actual 1695 mezzotint from a collection by Holman [33] is shown in Figure 4.5. The patterns are not as structured as that of a periodic screen, but do *not* have frequency components which go to zero as in random dither. The ancient mezzotint engravers would probably be outraged at the association. A true mezzotint beautifully renders delicate shades of gray without the graininess seen in white noise.

4.2 Hexagonal Grids

Hexagonal radially averaged power spectra exhibit the same well behaved features as in the rectangular case. A random dithered gray scale ramp and scanned image are shown in Figures 4.6 and 4.7 for regular hexagonal grids of the first kind. The radially average power spectrum and anisotropy have been observed for several gray levels. One representative sample at $g = \frac{1}{8}$ is shown in Figure 4.8.

Figure 4.5: Detail from a 1695 Mezzotint.

Figure 4.6: Hexagonal Random Dither of a Gray Scale Ramp.

Figure 4.7: Hexagonal Random Dither of a Scanned Picture.

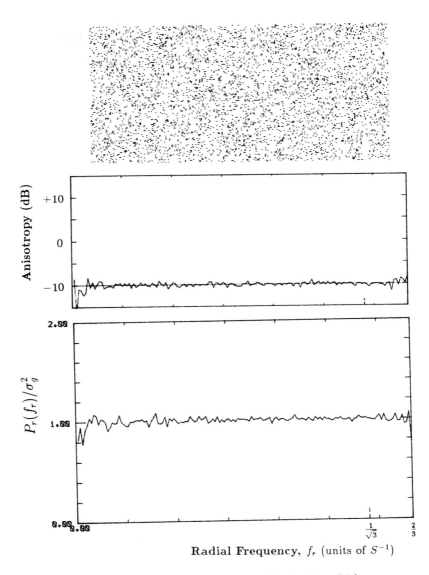

Figure 4.8: Hexagonal Radial Spectrum for Random Dither.
$\boxed{g = \frac{1}{8}}$, $\sigma_g^2 = \frac{7}{64} \approx .1094$

Chapter 5

Clustered-Dot Ordered Dither

Halftoning by ordered dither is the topic of this and the next two chapters. An ordered dither algorithm generates a binary halftone image by comparing pixels from an original continuous-tone image to a threshold value from a deterministic, periodic array. The thresholds are "ordered" rather than "random". Ordered dither is a point operation, that is, the output depends only on the state of the current pixel. Once a suitable threshold array is determined, the extreme simplicity of implementing halftoning by ordered dither makes it an important practical design candidate; it is for this reason that three chapters are dedicated to it.

The approach should not be confused with a method called Pulse-Surface-Area Modulation (PSAM) which maps an input pixel into a cell with a black to white distribution proportional to the original gray value. This is probably the earliest approach to computer halftoning [28,38,55] where variations ranged from overprinting of characters, use of special "gray-scale fonts", and arrangements of dots in fixed size cells. In the latter case, the mapping from input to output pixels is not one to one; an implied scale change occurs. There are no advantages of halftoning with PSAM, especially compared to the quality achieved with ordered dither.

Ordered dithering techniques can be divided into two classes by the nature of the "dots" produced, clustered and dispersed. In general, dispersed-dot ordered dither (to be addressed in Chapters 6 and 7) is preferred, but clustered-dot dither must be used for those binary display devices whose physical reconstruction functions cannot properly display isolated pixels. An example of such an electronic display was seen in section 2.2.

The most well known example of a system that must use clustered-dot halftones is the printing process. It is the printing process that gave us the word "halftoning". Clustered-dot ordered dither is by far the most widely used halftoning technique, electronic or otherwise, and is in fact the type assumed when the word *halftoning* is used without further qualification. It is interesting to note that to date, it is the only type of halftoning that is supported by a popular page description language [1].

5.1   The Classical Screen

The art of printing halftones is over a century old [33], and has experienced almost no change since it first came into practice. Besides being old, the printing industry today is one of the largest in the country, three times as large as the semiconductor industry. The vast majority of pictures produced every day are made on printing presses [68]. We should take serious note of those practices which have survived the decades.

Around 1850, the feasibility of the process was demonstrated by photographing an image through a loosely woven fabric or "screen" placed some distance from the focal plane. It came into practical use in the 1890's when the halftone screen became commercially available, consisting of two ruled glass plates cemented together. The quality of the resulting image depended a great deal on the skill of the printer, since the optimum distance at which the screen should be placed from the focal plane depended on a complex combination of parameters.

The only significant advance occurred in the 1940's with the introduction of the contact screen, a film bearing a properly exposed light distribution of a conventional screen. Not only did the contact screen eliminate cumbersome geometric considerations, but eliminated diffraction effects. Halftone dots produced with such a screen do not only vary in size but also in shape, thus allowing more high frequency detail to be produced at a given screen period.

This screen is sometimes called the graphic arts or printer's screen; in this book it is referred to as the "classical" screen. Today this ancient screen is the best available for preparing pictures that are to be reproduced by means of the printing process. In fact several photos in this book have been printed using the classical screen, such as those seen in section 2.2. Even in this chapter, the three enlarged photographs of printed classical screens in Figure 5.1, are themselves printed with this screen! (It should be pointed out that the resolution required to generate these images digitally would be at least 12 times greater than that used in digitally produced examples in this book.)

Naturally, for low visibility of the screen, as small a period as possible is desired. But this is limited by factors such as the viscosity of the ink, paper coarseness, and the minimum printing plate area that will hold a dot of ink. Details of the use of this screen in the printing process are reviewed by Schreiber [70, ch. 6] and Yule [90]. Roetling [61] has quantified the number of perceptually detectable gray levels possible for this screen as a function of spatial frequency.

5.1.1 Orientation Sensitivity

The halftoning process shares many of the perceptual concerns common with most image processing problems. However, general texts such as by Graham [27] or Cornsweet [16] omit one issue that is specifically important in halftoning.

Figure 5.1 shows the detail of actual printed halftones produced with a (a) 125 lpi screen in 1902, (b) 150 lpi screen in 1916, and (c) 88 lpi screen in 1984. These pictures reveal that since the dawn of halftoning it was recognized that the resulting images "looked better" if the screen was oriented at a 45 degree angle.

We can demonstrate the lack of symmetry in the frequency response of the visual system by looking at Figure 5.2. Although this phenomenon was known and accounted for in the earliest halftone screen orientations, it wasn't experimentally measured until several decades later [13,31,81].

Figure 5.3, qualitatively shows our spatial frequency sensitivity as a function of orientation. It is based on data presented by Taylor [81] who measured the detectability of a fine grating as orientation was varied. Note the sharp cusps at the horizontal and vertical orientations.

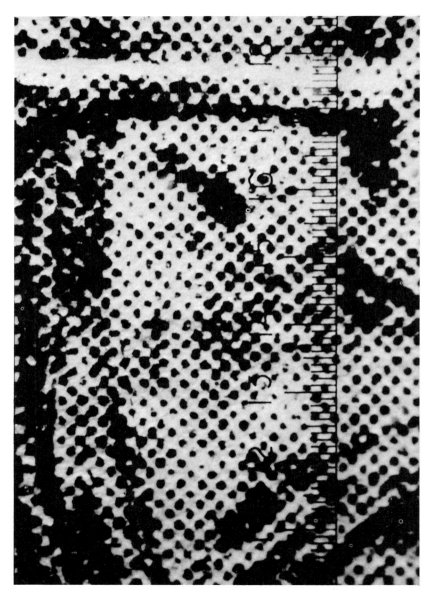

Figure 5.1: Microphotographs of printed classical halftones.
(a) Detail from a book published in 1902 [41].
(Numbered reticle marks are millimeters.)

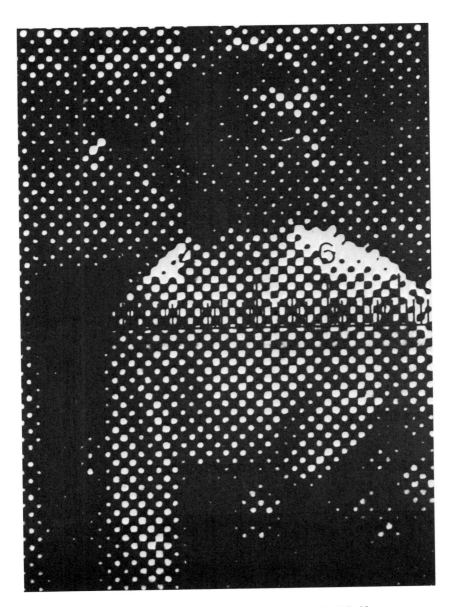

Figure 5.1: Microphotographs of printed classical halftones.
(b) Detail from a book published in 1916 [46].
(Numbered reticle marks are millimeters.)

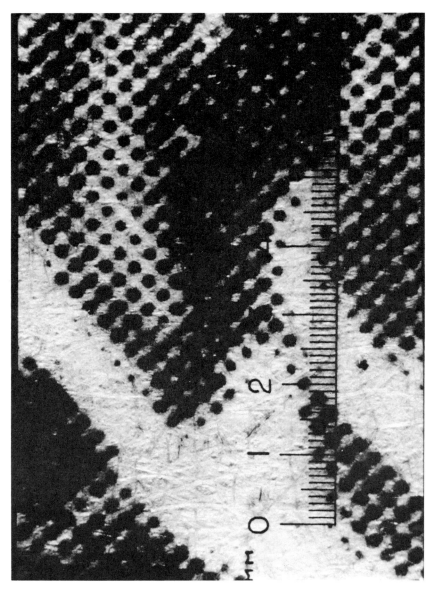

Figure 5.1:   Microphotographs of printed classical halftones.
(c) Detail from a contemporary newspaper.
(Numbered reticle marks are millimeters.)

(a) Screen at 45 degree angle.

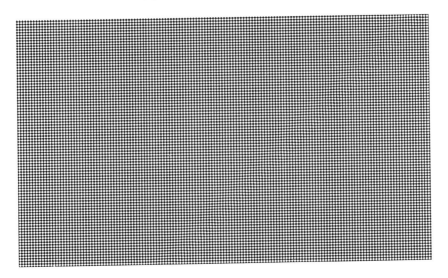

(b) Horizontal screen.

Figure 5.2: Orientation Perception of a 33 lpi screen.
(9 cycles/degree at 40 cm)
Horizontal and vertical lines are seen in the horizontal screen.

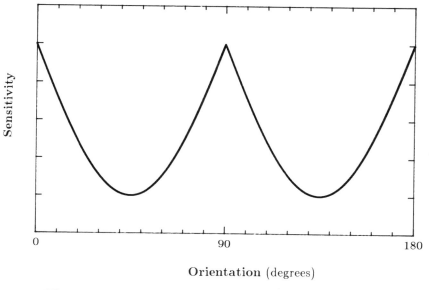

Orientation (degrees)

Figure 5.3: Spatial Frequency Sensitivity vs. Orientation.

Also, this disparity in sensitivity to oblique and horizontal gratings increases with spatial frequency [13].

5.2   Rectangular Grids

Several major classes of clustered-dot halftones will now be demonstrated digitally, first for rectangular, then for hexagonal grids. In all cases, the threshold array, halftoned gray scale ramp, halftoned scanned picture, and exposure plot of the composite Fourier transform will be shown.

5.2.1   Classical Screen at $45°$

Three sizes of this, the most popular halftone screen will be examined in this section. Generating the ordered threshold array for this screen is very simple. At middle gray ($g = \frac{1}{2}$) the output plane will be covered by a checkerboard pattern of alternating black and white squares of size $M \times M$ pixels. For gray levels lighter than $g = \frac{1}{2}$, the black squares must diminish is size, and for darker gray levels, the white squares contract as black pixels spiral in.

The threshold array will have $Z = 2M^2$ elements, and thus $2M^2 + 1$ gray levels can be represented. The tradeoff in the selection of any ordered dither threshold array size is always between the visibility of the low frequency due to the period or size of the array and the number of gray levels which can be rendered. Threshold arrays for $M = 3$, 4, and 8 are shown in Figure 5.4.

Halftoning with threshold arrays such as these simulate the contact screen in that they allow high contrast, high frequency detail to punch through the screen. Simulating the pre-1940's screen would require averaging or low-pass filtering the image data in the threshold period prior to screening.

Note that the period is odd, and as is the convention in this text, the period is shown with nonunique edges (see Figure 3.1) to better illustrate the nature of the periodicity. Also, borders between the dark and light halves of the threshold array have been drawn. In this and all other illustrations of threshold arrays, the values shown are actually the order in which the thresholds are arranged rather that their absolute value.

The effect of halftoning a gray scale ramp is shown in Figure 5.5, a scanned image in Figure 5.6, and a synthesized image in Figure 5.7. These illustrations underline the extremely low resolution used to display results in this book. The extremely coarse screen with $M = 8$, would look quite reasonable if the longer dimension of the images shown were shrunk to about 1.5 cm at normal viewing distance, or if the figure was viewed at a distance of 5 meters.

Exposure plots of composite Fourier transforms (see section 3.1) are shown in Figure 5.8. At the bottom of each exposure plot, a scale defining the actual dimensions (in cycles/unit-length) of the plot in terms of the original sample period, S, is provided. The shape of the baseband reflects the shape of the pixel, which remains constant (square) in this case. Because the spatial period was odd, the frequency samples are arranged in an odd fashion. Also, for all period sizes, the distribution of energy in the frequency domain is concentrated at the low frequency center in a symmetric fashion.

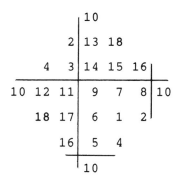

(a) $M = 3$ (19 levels of gray).

```
                    19
               10 | 28  29
            2  11 | 27  32  31
         7  8  15 | 24  26  25  18
  19  21 20 17 | 14 12 13 16 | 19
     29 30 23 |  5  4  3 10
        31 22 |  6  1  2
           18 |  9  7
              | 19
```

(b) $M = 4$ (33 levels of gray).

Figure 5.4: Threshold arrays for 45° Classical Screens.

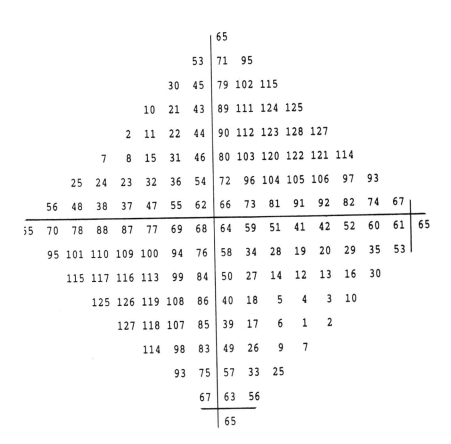

(c) $M = 8$ (129 levels of gray).

Figure 5.4: Threshold arrays for $45°$ Classical Screens (continued).

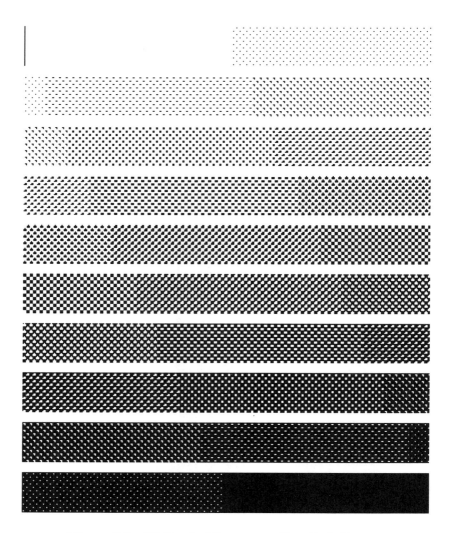

Figure 5.5: 45° Classical Screen on a Gray Scale Ramp.
(a) 19 levels of gray, $M = 3$.

Figure 5.5: 45° Classical Screen on a Gray Scale Ramp.
(b) 33 levels of gray, $M = 4$.

Figure 5.5: 45° Classical Screen on a Gray Scale Ramp.
(c) 129 levels of gray, $M = 8$.

Figure 5.6: 45° Classical Screen on a Scanned Picture.
(a) 19 levels of gray, $M = 3$.

Figure 5.6:   45° Classical Screen on a Scanned Picture.
(b) 33 levels of gray, $M = 4$.

Figure 5.6: 45° Classical Screen on a Scanned Picture.
(c) 129 levels of gray, $M = 8$.

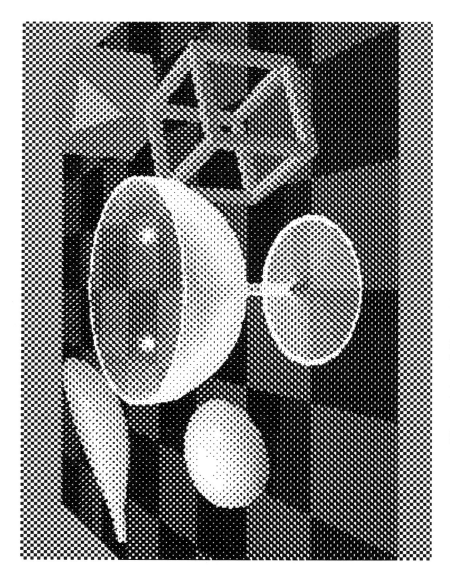

Figure 5.7: 45° Classical Screen on a Synthesized Image. (a) 19 levels of gray, $M = 3$.

Figure 5.7: 45° Classical Screen on a Synthesized Image. (b) 129 levels of gray, $M = 8$.

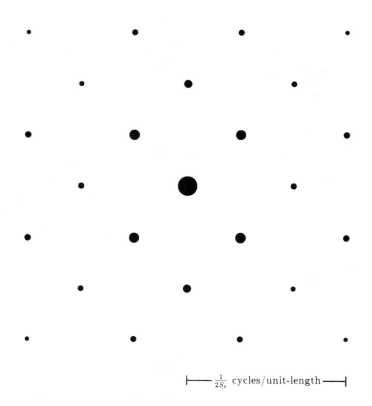

$\vdash\!\!\!-\!\!\!- \frac{1}{2S_r}$ cycles/unit-length $\!\!-\!\!\!\dashv$

Figure 5.8: Composite Fourier Transform of the 45° Classical Screen. (a) Average of 19 patterns, $M = 3$.

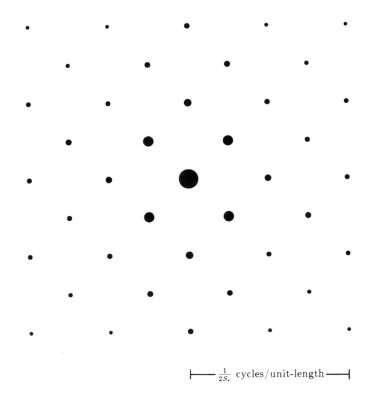

$\vdash$———$\frac{1}{2S_r}$ cycles/unit-length———$\dashv$

Figure 5.8: Composite Fourier Transform of the 45° Classical Screen. (b) Average of 33 patterns, $M = 4$.

$\longmapsto \frac{1}{2S_r}$ cycles/unit-length $\longmapsto$

Figure 5.8: Composite Fourier Transform of the 45° Classical Screen. (c) 129 levels of gray, $M = 8$.

5.2.2 Classical Screen at $0°$

Because of the sharp minimum in perceptual sensitivity for spatial frequencies oriented at $45°$ (or $135°$) from horizontal, there is no reason to generate screens at any other angle for binary displays. Printing color images, however, presents a new problem.

Color pictures require an overlaying binary image for each of three inks to represent a reasonable range of the color gamut. In the case of hard copy, a fourth overlaying image, black, is usually needed to increase the maximum density achievable, eliminate chromatic errors and the need for critical ink balance across the complete range of neutral grays, and to reduce cost by replacing the use of the three more expensive inks for one less expensive ink in regions of neutral gray.

Overlaying identical screens with a fixed phase shift would not be a problem if the output display was perfectly noiseless. However, even an extremely small but regular variation in dot position due to the Position Noise, $\delta(\mathbf{x} - \vec{\epsilon}(\mathbf{x}))$, in the Physical Reconstruction Function (Figure 2.1, page 16) common in real printers produce an overwhelming artifact, moiré patterns resulting from the beat frequencies between the periodic screens. When the overlaid screens are of different component colors, the effect is manifested as color shifts.

This problem has been well studied and is minimized in practice by simply orienting the multiple screens at different angles, usually about $15°$ apart (see [90, ch. 13]). Holladay [32] has presented a method for digitally generating so called "rational screens"—the classical screen at angles which have tangents that are ratios of relatively small integers. Techniques have also been invented [23,64] to produce such screens at *any* angle by "dithering" between angles with rational tangents. In this section, an example of a different screen angle is demonstrated at $0°$.

Figure 5.9 shows the 36 element threshold array used. Note that the period is even, and that the border between the dark and light halves of the screen is as indicated. The effect on a gray scale ramp and scanned picture are shown in Figure 5.10 and 5.11. The rectangular arrangement of frequency samples seen in the exposure plot (Figure 5.12) reflect the fact that the spatial period is even. As in the $45°$ example, energy is symmetrically concentrated at the low frequency center.

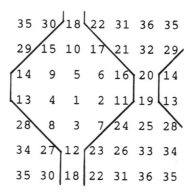

Figure 5.9: Threshold array for a 0° Classical Screen.

5.2.3 The Spiral-Dot and Line Screens

A wide range of special effect clustered-dot screens are available in the graphic arts industry and all are easily simulated digitally. Graphics systems are available which do this in hardware. In this section, two such screens are examined.

Figure 5.13 shows the threshold arrays for the spiral-dot and line screens. The line screen clusters pixels about horizontal lines, while the spiral-dot screen is essentially half of the classical screen, with dark squares growing to fill the plane without the alternating light squares.

The gray scale ramp and scanned image are displayed in Figures 5.14 and 5.15 for the spiral-dot screen and Figures 5.17 and 5.18 for the line screen. The exposure plot of the 5×5 element spiral-dot screen (Figure 5.16) is quite similar to that of the 0° classical screen. This is the only even period rectangular threshold array with an odd number of elements on a side in this book. The horizontal and vertical periods are 5 units. For this reason, no frequency energy can exist at the edges of the baseband, at $f_1 = \frac{1}{2S}$ or $f_2 = \frac{1}{2L}$.   The lack of energy there is not an issue for a clustered-dot screen, but would be for dispersed-dot ordered dither (investigated in Chapter 6), the success of which depends on the concentration of energy at high frequencies.

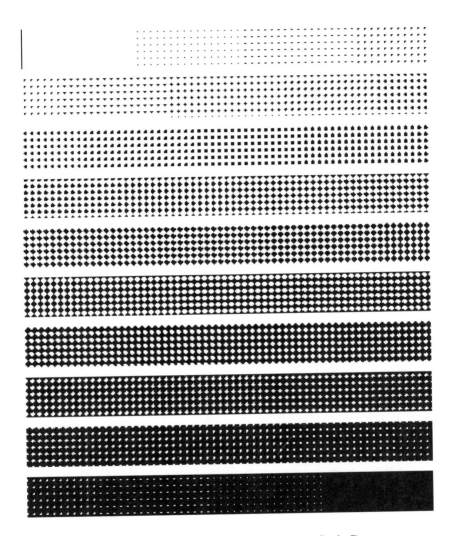

Figure 5.10: 0° Classical Screen on a Gray Scale Ramp.
(37 levels of gray.)

Figure 5.11: 0° Classical Screen on a Scanned Picture.
(37 levels of gray.)

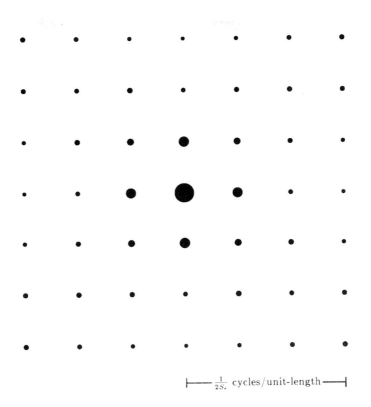

$\longmapsto \frac{1}{2S_r}$ cycles/unit-length $\longmapsto$

Figure 5.12: Composite Fourier Transform of the 0° Classical Screen. (Average of 37 patterns.)

```
21  22  23  24  25  21
20 ┌ 7   8   9  10 ┐20
19 │ 6   1   2  11 │19
18 └ 5   4   3  12 ┘18
17  16  15  14  13  17
21  22  23  24  25  21
```

Spiral-Dot Screen

Line Screen

```
36  34  32  31  33  35  36
24  22  20  19  21  23  24
────────────────────────────
12  10   8   7   9  11  12
 6   4   2   1   3   5   6
────────────────────────────
18  16  14  13  15  17  18
30  28  26  25  27  29  30
36  34  32  31  33  35  36
```

Figure 5.13: Threshold arrays for Spiral-Dot and Line Screens.

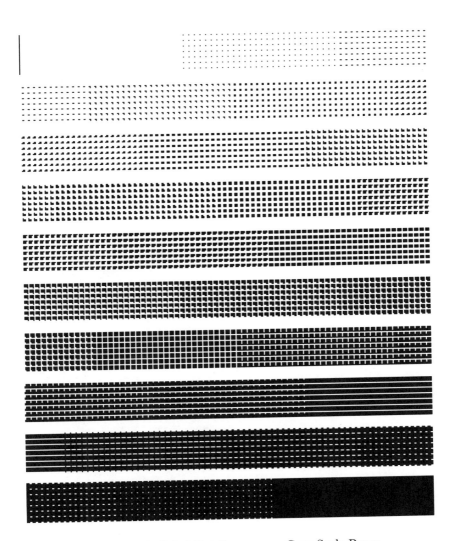

Figure 5.14: Spiral-Dot Screen on a Gray Scale Ramp.
(26 levels of gray.)

Figure 5.15: Spiral-Dot Screen on a Scanned Picture.
(26 levels of gray.)

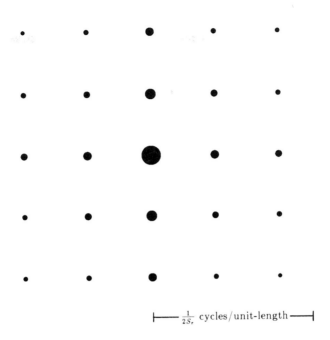

$$\vdash\!\!\!\!\!-\!\!\!\!-\tfrac{1}{2S_r} \text{ cycles/unit-length} -\!\!\!\!\dashv$$

Figure 5.16: Composite Fourier Transform of the Spiral-Dot Screen.
(Average of 26 patterns.)

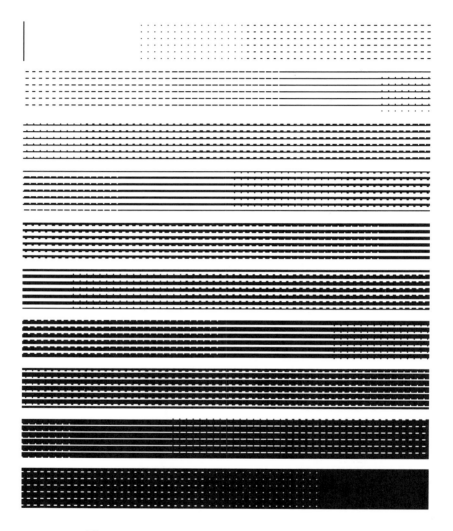

Figure 5.17: Line Screen on a Gray Scale Ramp.
(37 levels of gray.)

Figure 5.18: Line Screen on a Scanned Picture.
(37 levels of gray.)

The exposure plot of the 36 element line screen in Figure 5.19 reveals a strong concentration along the vertical frequency axis. This is due to the dominant vertical frequency resulting from the preponderance of horizontal lines. These vertical frequencies are separated by the inverse of the fundamental vertical spatial period, $\frac{1}{6L}$.

5.2.4   Asymmetric Correction

Thus far, only regular grids have been considered. The problem of correcting a threshold array for an asymmetric grid is much more complicated for dispersed-dot ordered dither and well be treated in depth in Chapter 7. The problem is comparatively simple for clustered-dot ordered dither, and thus only one representative example will be presented in this section.

Clustered-dot threshold arrays can be thought of as samples of slowly varying continuous threshold functions. Changing the aspect ratio of the grid requires sampling the threshold function with a different aspect ratio.

The example to be used here will be the 45° classical screen with $M = 8$ (examined in section 5.2.1) on a grid with aspect ratio, $\alpha = \frac{1}{2}$. Figure 5.20 displays the unwanted elongation that will occur if the threshold array of Figure 5.4(c) is used directly to a scanned image on such a grid.

A corrected threshold array is given in Figure 5.21, along with the results on the gray scale ramp and scanned picture with $\alpha = \frac{1}{2}$ in Figures 5.22 and 5.23. Figure 5.24 compares the exposure plots due to the (a) uncorrected threshold array and (b) corrected threshold array. It is first observed that the vertical size of the baseband has been halved because the line period, L, of the spatial pixel was doubled. Figure 5.24(a) is simply a squashed version of Figure 5.8(c). The frequency samples at lower frequency on the vertical frequency axis are evident in the longer vertical period in the spatial domain picture of Figure 5.20.

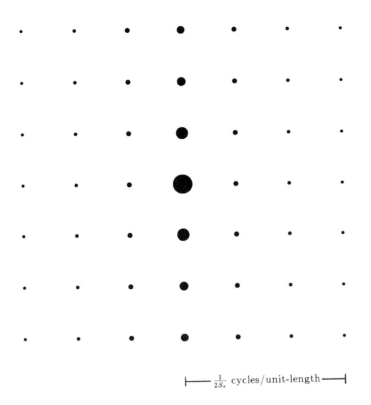

$\vdash\!\!\!\!\!\longrightarrow \frac{1}{2S_r}$ cycles/unit-length $\longrightarrow\!\!\!\!\!\dashv$

Figure 5.19: Composite Fourier Transform of the Line Screen. (Average of 37 patterns.)

Figure 5.20: Uncorrected Classical Screen on a Scanned Picture. Threshold array from Figure 5.4(c), $\alpha = \frac{1}{2}$.

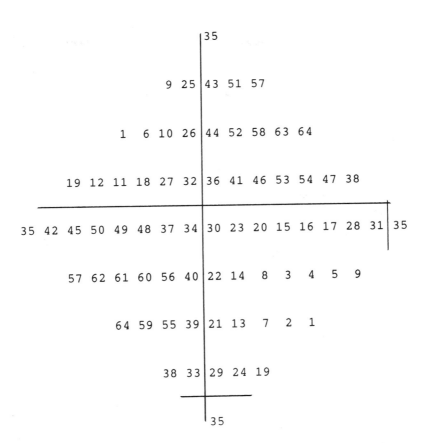

Figure 5.21: Corrected Threshold Array for a grid with $\alpha = \frac{1}{2}$.

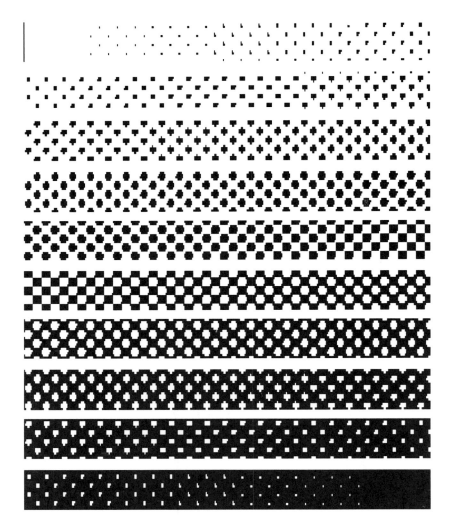

Figure 5.22: Corrected Classical Screen on a Gray Scale Ramp, $\alpha = \frac{1}{2}$. (65 levels of gray.)

Figure 5.23: Corrected Classical Screen on a Scanned Picture, $\alpha = \frac{1}{2}$.
(65 levels of gray.)

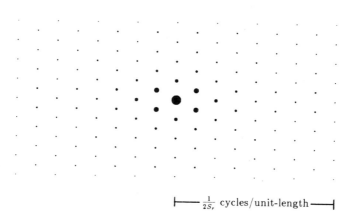

$\longmapsto \frac{1}{2S_r}$ cycles/unit-length $\longmapsto$

(a) Uncorrected Threshold Array of Figure 5.4(c).

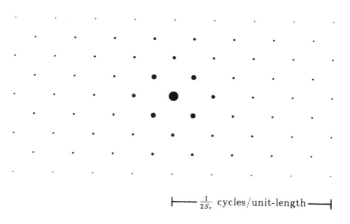

$\longmapsto \frac{1}{2S_r}$ cycles/unit-length $\longmapsto$

(b) Corrected Threshold Array of Figure 5.21.

Figure 5.24: Composite Fourier Transforms with $\alpha = \frac{1}{2}$.

5.3 Hexagonal Grids

Two techniques for rendering clustered dot screens on regular hexagonal grids will now be presented.

5.3.1 Hexagonal Version of the Classical Screen

The desirable properties of the rectangular classical screen which should be preserved in developing a hexagonal version are as follows:

1. Clusters should be as symmetrically distributed as possible.

2. There should exist one connected cluster per period.

3. Gray scale symmetry should be exhibited; that is, the pattern generated for gray level, g, should be the inverse of the pattern generated for $1 - g$.

While "gray scale symmetry" is not necessarily a desirable property of the classical screen in the traditional printing process, variations in the Physical Reconstruction Functions of electronic display devices make it important; for example, it may be used on laser printers that generate the latent charge image onto the drum by either erasing white or writing black. Isolated white pixels are larger than isolated black pixels, or vice versa, depending on which type of printer is used. A grey-scale-symmetric screen can be used on either, accompanied by an appropriate tone-scale adjustment.

Figure 5.25 illustrates the tristate ordering scheme proposed for generating such a hexagonal screen. The base period consists of three hexagons. Thresholds in the "light" and "dark" hexagons spiral in and out in the same way as in the light and dark squares of the rectangular classical screen. However, to maintain the above three properties, the hexagon associated with the middle third has threshold ordered in the zigzag manner shown.

A 27 element screen generated in this way is shown in Figure 5.26. The period is displayed in the standard hexagonal shape. The outlines show the borders between the three subhexagons. The resulting gray scale ramp and scanned picture are shown in Figures 5.27 and 5.28. These 28 gray level hexagonal patterns can best be compared against the 33 gray level rectangular patterns of section 5.2.1 for $M = 4$. The composite Fourier transform in Figure 5.29 displays the expected symmetric concentration of energy about the zero frequency term.

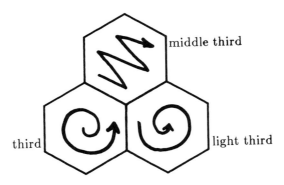

Figure 5.25: Tristate Ordering Scheme for the Hexagonal Classical Screen.

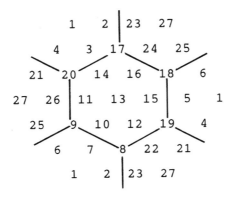

Figure 5.26: Threshold array for a 27 element Hexagonal Classical Screen.

Figure 5.27: Hexagonal Classical Screen on a Gray Scale Ramp.
(28 levels of gray.)

Figure 5.28: Hexagonal Classical Screen on a Scanned Picture.
(28 levels of gray.)

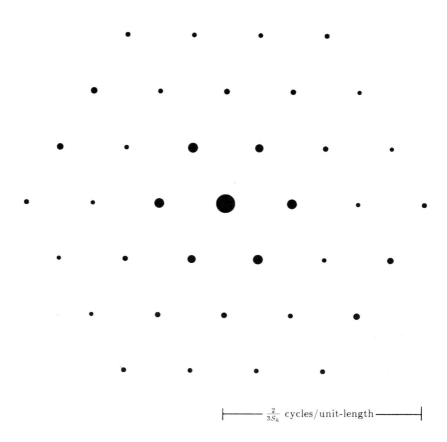

$\vdash\!\!\!\!\!-\!\!\!-\!\!\!-\!\!\!-\!\!\!-\!\!\! \frac{2}{3S_h}$ cycles/unit-length $-\!\!\!-\!\!\!-\!\!\!\dashv$

Figure 5.29: Composite Fourier Transform
of the Hexagonal Classical Screen.
(Average of 28 patterns.)

Figure 5.30: Threshold array for a Hexagonal Spiral-Dot Screen. (28 levels of gray.)

5.3.2 Spiral-Dot Screen

The images generated here are simply hexagonal versions of the rectangular spiral-dot screen of section 5.2.3. Various sizes of such a screen were proposed in a study by Chao [14, pp. 45-47]. However, the threshold arrays shown in that work would not tile the plane in a regular hexagonal manner.

A 27 element spiral-dot threshold array is shown in Figure 5.30 along with its results on the gray scale ramp (Figure 5.31) and scanned picture (Figure 5.32). Again, the composite Fourier transform, shown in Figure 5.33, exhibits the concentration of energy near the zero frequency term.

A hexagonal spiral-dot screen shares the same shortcomings as in the rectangular case (see pages 105 and 106) in that it fails to meet one one of the conditions enumerated on page 117; it does not possess gray scale symmetry. In light regions, black pixels form well nucleated clusters, but in dark regions, the white pixels are not well nucleated. One must keep in mind that the robust clustering of both white and black pixels is the reason for settling for a clustered-dot screen. If this this property is not needed, then the visually superior dispersed-dot patterns presented in the next section for either rectangular or hexagonal grids are the ones of choice for ordered dither.

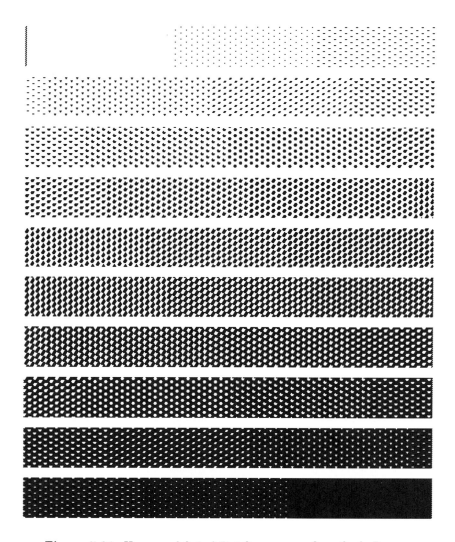

Figure 5.31: Hexagonal Spiral-Dot Screen on a Gray Scale Ramp.
(28 levels of gray.)

Figure 5.32: Hexagonal Spiral-Dot Screen on a Scanned Picture.
(28 levels of gray.)

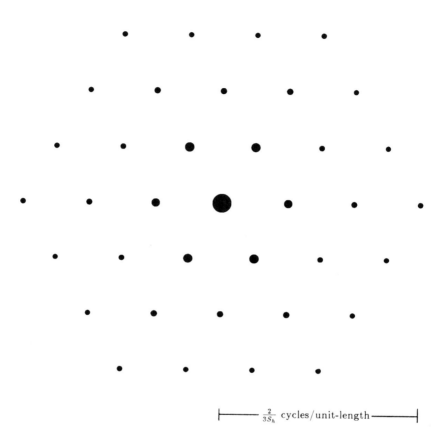

$\longmapsto \quad \frac{2}{3S_h} \text{ cycles/unit-length} \longrightarrow\!|$

Figure 5.33: Composite Fourier Transform
of the Hexagonal Spiral-Dot Screen.
(Average of 28 patterns.)

Chapter 6

Dispersed-Dot Ordered Dither

When the image is to be produced on a device that can faithfully display every binary pixel, the preferred choice of ordered dither is one that generates dispersed rather than clustered dots. Dispersed-dot threshold arrays yield high frequency fidelity and illusions of constant gray regions better than do clustered-dot arrays of the same resolution and period.

The design of dispersed-dot ordered dither threshold arrays has been studied on square grids by Limb [42], Lippel [43,44], and most notably by Bayer. Halftoning with a particular homogeneous threshold matrix has become known as "Bayer's dither" after his famous 1973 proof [10] of optimality with respect to minimizing low frequency texture. In fact, in a wide sense, "ordered dither" has come to mean Bayer's dither.

In this chapter, the Method of Recursive Tessellation [83], a technique for generating optimally homogeneous ordered dither threshold arrays on both square and hexagonal lattices is introduced and examined in the frequency domain. Not only does this method generalize ordered dither for hexagonal grids, but for rectangular grids as well in that it applies to both even and odd period threshold arrays. It will be shown that only odd period arrays should ever be used for rectangular grids.

6.1   Method of Recursive Tessellation

The Latin word *tesselare* means "to pave with tiles". In this section, a method for deriving optimally homogeneous threshold arrays for regularly shaped periods of grid points (tiles) on regular grids is presented.

6.1.1   Tessellating Regular Grids

Rectangular grids are regular for $\alpha = 1$ only. Semiregular hexagonal grids are regular at three aspect ratios, $\alpha = \frac{2}{\sqrt{3}}$ (regular hexagon of the first kind), $\alpha = 2\sqrt{3}$ (regular hexagon of the second kind), and $\alpha = 2$ (square). Two cases will be worked out in detail, rectangular with $\alpha = 1$ and hexagonal with $\alpha = \frac{2}{\sqrt{3}}$, the default "rectangular" and "hexagonal" cases. The results can later be applied to the remaining two cases by rotation. Rotate the hexagonal result by 90° (or 30°) for hexagonal grids of the second kind ($\alpha = 2\sqrt{3}$); for hexagonal grids with $\alpha = 2$, rotate the rectangular result by 45°.

The fundamental period or tile of the threshold array can have an integral power of 2 elements for the rectangular case, and an integral power of 3 elements for the hexagonal case. The power of 2 or 3 is defined as the *order*, η, of the array. Figure 6.1 shows the first 8 orders of rectangular tiles and Figure 6.2 shows the first 5 hexagonal orders. The labeling of families of periods as "even" or "odd" in section 3.1 is consistent with periods with even and odd orders as defined here.

Once a grid and period order is selected, all of two-space is tiled.

6.1.2   Generation of Threshold Arrays

The goal is to order the samples from 1 to 2^η for the rectangular case or 1 to 3^η for the hexagonal case in such a way that as each successive position is numbered (turned on), the total two-dimensional ensemble of "on" positions is as homogeneously arranged as possible. When used as threshold arrays, the corresponding arrangement of output binary dots will be dispersed as homogeneously as possible for each gray level to be simulated.

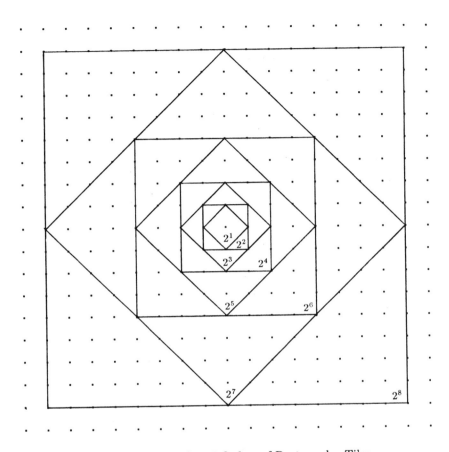

Figure 6.1: The first 8 Orders of Rectangular Tiles. The number of elements unique to each tile is 2^7.

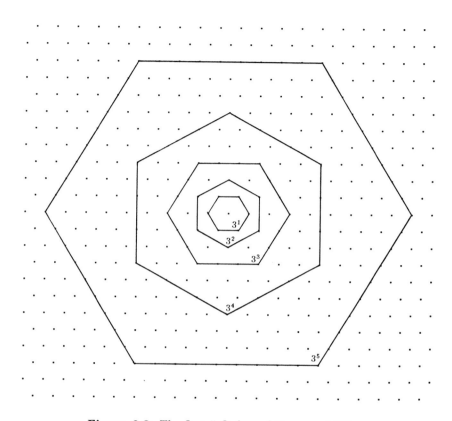

Figure 6.2: The first 5 Orders of Hexagonal Tiles.
The number of elements unique to each tile is 3^7.

The algorithm for generating the threshold array pivots on the fact that starting with a tile of a given order, η, the center (prime) point and all the vertices can act as center points for a retiling with tiles of order $\eta - 1$. The vertices of these tiles can further act as center points for another retiling of order $\eta - 2$, and so on. All of the vertices at each stage of this recursive tessellation are numbered before the next tessellation takes place. Breaking down the plane in this way provides a mechanism for locating the family of points that are exactly in the center of the voids between points of the higher order families.

The ordering of the family of points within a given stage or "subtessellation", is governed by an offset vector between the central (prime) point and any one of the vertices of its circumscribing tile. Figure 6.3 illustrates the process for a fourth order rectangular array. For an array of order η, there are η stages or subtessellations. At each stage, i, 2^i points are numbered by placing the tail of an offset vector on the center of each tile to locate one vertex in that tile. The offset vector can be initially oriented to point at any vertex, but must not change orientation within a given stage. Each vertex located with the offset vector is assigned a number equal to that of the center point plus the number of points assigned in the preceding stage, or 2^{i-1}.

The same procedure applies to the hexagonal case, except that each stage requires 2 passes, and the total number of points assigned in each stage is 3^i. The 2 page example illustrated in Figure 6.4 shows the method of recursive tessellation for a hexagonal tile of order 3. Two passes are required for each of the 3 stages. Note that on each second pass, the tail of the offset vector is placed on points assigned in the first pass instead of on tile centers.

Figure 6.5 (pages 135 through 137) shows the resulting rectangular threshold arrays for $\eta_r = 1$ through 8. These arrays are the same as described by Bayer [10]. The first 5 orders of hexagonal threshold arrays are shown in Figure 6.6 (pages 138 and 139). Recall that because the array edges are shared when tiling the plane, two edges of each rectangular tile and three edges of each hexagonal tile are not unique (Figure 3.1, page 43).

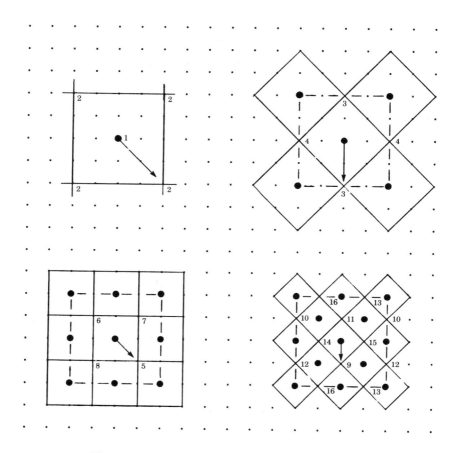

Figure 6.3: The 4 stages of Recursive Tessellation
of a fourth order rectangular tile.

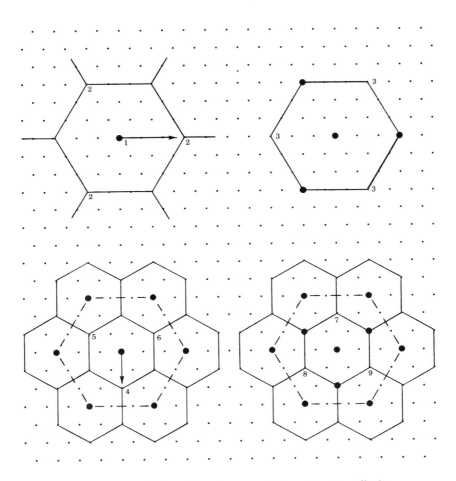

Figure 6.4: (a) The first 2 stages of Recursive Tessellation
of a third order hexagonal tile.
Two passes are required for each stage.

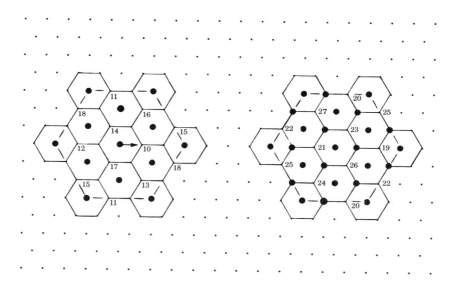

Figure 6.4:   (b) The third stage of Recursive Tessellation
of a third order hexagonal tile.

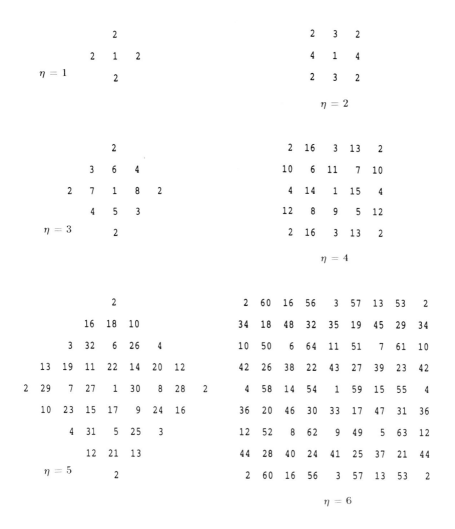

Figure 6.5: Rectangular Threshold Arrays.
The edges of each array are shared in the tiling.
(Continued on next two pages.)

```
                              2
                        60  66  34
                    16 124  18  98  10
                56  80  48  82  50  74  42
             3 120  32 112   6 114  26 106   4
        57  67  35  96  64  70  38  90  58  68  36
    13 121  19  99  11 128  22 102  14 122  20 100  12
   53  77  45  83  51  75  43  86  54  78  46  84  52  76  44
2 117  29 109   7 115  27 107   1 118  30 110   8 116  28 108   2
   34  93  61  71  39  91  59  65  33  94  62  72  40  92  60
   10 125  23 103  15 123  17  97   9 126  24 104  16
       42  87  55  79  47  81  49  73  41  88  56
            4 119  31 111   5 113  25 105   3
               36  95  63  69  37  89  57
                   12 127  21 101  13
                       44  85  53
```

$\eta = 7$ 2

Figure 6.5: Rectangular Threshold Arrays (continued).

```
  2 236  60 220  16 232  56 216   3 233  57 217  13 229  53 213    2
130  66 188 124 144  80 184 120 131  67 185 121 141  77 181 117  130
 34 194  18 252  48 208  32 248  35 195  19 249  45 205  29 245   34
162  98 146  82 176 112 160  96 163  99 147  83 173 109 157  93  162
 10 226  50 210   6 240  64 224  11 227  51 211   7 237  61 221   10
138  74 178 114 134  70 192 128 139  75 179 115 135  71 189 125  138
 42 202  26 242  38 198  22 256  43 203  27 243  39 199  23 253   42
170 106 154  90 166 102 150  86 171 107 155  91 167 103 151  87  170
  4 234  58 218  14 230  54 214   1 235  59 219  15 231  55 215    4
132  68 186 122 142  78 182 118 129  65 187 123 143  79 183 119  132
 36 196  20 250  46 206  30 246  33 193  17 251  47 207  31 247   36
164 100 148  84 174 110 158  94 161  97 145  81 175 111 159  95  164
 12 228  52 212   8 238  62 222   9 225  49 209   5 239  63 223   12
140  76 180 116 136  72 190 126 137  73 177 113 133  69 191 127  140
 44 204  28 244  40 200  24 254  41 201  25 241  37 197  21 255   44
172 108 156  92 168 104 152  88 169 105 153  89 165 101 149  85  172
  2 236  60 220  16 232  56 216   3 233  57 217  13 229  53 213    2
```

$$\eta = 8$$

Figure 6.5: Rectangular Threshold Arrays (continued).

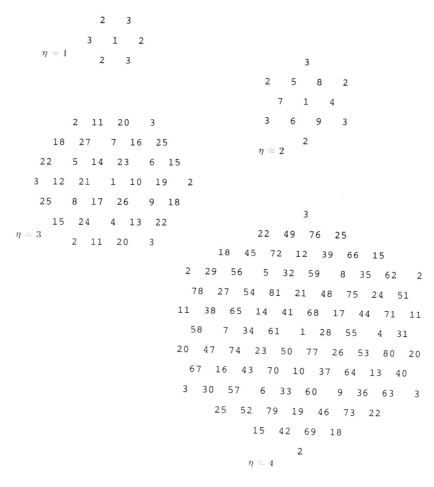

Figure 6.6: Hexagonal Threshold Arrays.
The edges of each array are shared in the tiling.
(Continued on next page.)

```
    2   83 164   11   92 173   20 101 182    3
     150 231   78 159 240   58 139 220   67 148 229
      208   29 110 191   38 119 200   47 128 209   30 111
       18   99 180   27 108 189    7   88 169   16   97 178   25
        154 235   56 137 218   65 146 227   74 155 236   57 138 219
        198   45 126 207   54 135 216   34 115 196   43 124 205   52 133
        22 103 184    5   86 167   14   95 176   23 104 185    6   87 168   15
       144 225   72 153 234   81 162 243   61 142 223   70 151 232   79 160 241
      202   49 130 211   32 113 194   41 122 203   50 131 212   33 114 195   42 123
      3   84 165   12   93 174   21 102 183    1   82 163   10   91 172   19 100 181    2
      229   76 157 238   59 140 221   68 149 230   77 158 239   60 141 222   69 150
       111 192   39 120 201   48 129 210   28 109 190   37 118 199   46 127 208
        25 106 187    8   89 170   17   98 179   26 107 188    9   90 171   18
         219   66 147 228   75 156 237   55 136 217   64 145 226   73 154
          133 214   35 116 197   44 125 206   53 134 215   36 117 198
           15   96 177   24 105 186    4   85 166   13   94 175   22
            241   62 143 224   71 152 233   80 161 242   63 144
             123 204   51 132 213   31 112 193   40 121 202
    2   83 164   11   92 173   20 101 182    3
```

$\eta = 5$

Figure 6.6: Hexagonal Threshold Arrays (continued).

As a practical matter, for operating on rectangularly shaped images, 2 periods of these arrays have to be packed together to form rectangularly periodic matrices. The exception, of course, is rectangular threshold arrays with η even, which are already rectangularly periodic. Figure 6.7 shows how this packing is done for odd ordered rectangular and hexagonal arrays, and Figure 6.8 illustrates this for even ordered hexagonal arrays. Note that rectangular tessellation of these packed matrices preserves the periodicity of the originally shaped tiles.

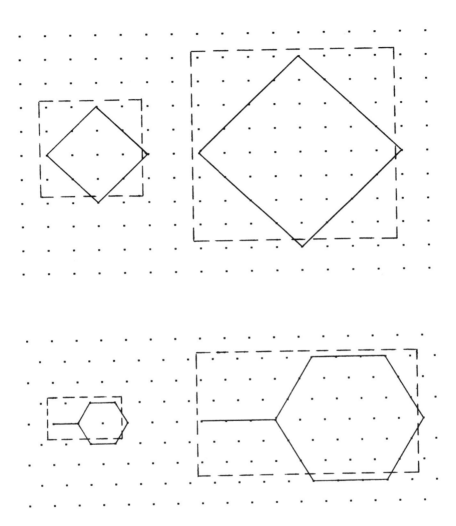

Figure 6.7: Packing two odd ordered periods for rectangular storage.

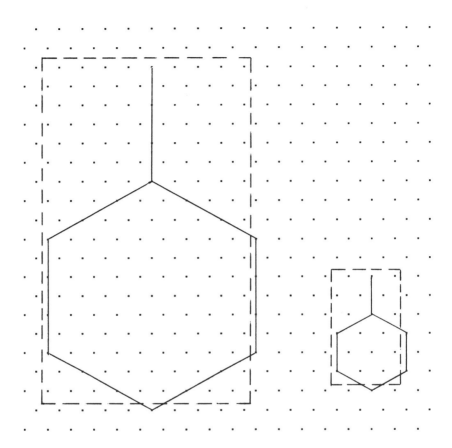

Figure 6.8: Packing two even ordered periods for rectangular storage.

6.2 Examples for Regular Grids

To assure a fair comparison, each of the images shown in this section has an equal number of samples per unit area. Figure 6.9 (pages 144 through 151) shows the result of halftoning a wrapped one-dimensional gray scale ramp on a rectangular grid with each of the first 8 orders. As usual, the beginning of the ramp is marked with a black line. In each case, $2^{\eta_r} + 1$ gray levels (including white) are simulated. Similarly, the first 5 hexagonal orders are shown in Figure 6.10 (pages 152 through 156). Figure 6.11 (pages 158 through 162) and Figure 6.13 (pages 165 through 168) illustrate the result of halftoning a scanned image. Two examples of rectangular ordered dither of the synthesize image is shown in Figure 6.12.

These examples demonstrate the tradeoff between different threshold array orders. Note that "zeroth order" halftoning would correspond to using a single fixed threshold as in Figure 1.1 (page 4). For small orders, spurious contours due to texture changes in the patterns result from the insufficient number of gray levels. Images produced with large order threshold arrays suffer from the appearance of low frequency patterns in areas of uniform gray. The choice of optimum order, η, depends on the resolution in cycles/degree with which the image is to be displayed.

In any case, the dominance of the basic period seen in clustered-dot ordered dither is gone. The images of Figure 6.11(e) (page 162) and Figure 6.12(b) (page 164) for $\eta_r = 7$ and the classical screens of Figure 5.6(c) (page 93) and Figure 5.7(b) (page 95) each have a period of *exactly* the same size ($Z = 128$) and shape!

Upon comparing rectangular and hexagonal images produced with comparable number of gray levels, the images on the hexagonal grid possess patterns that are less disturbing than those on the rectangular grid. As explained in section 5.1.1, human vision is more acute for horizontal and vertical orientations than for screens oriented at a 45° angle. The preponderance of horizontal and vertical patterns seen in rectangular ordered dither is not present in the hexagonal case, which probably explains its more pleasing appearance.

Figure 6.9: Rectangular Ordered Dither of a Gray Scale Ramp.
(a) 3 (or $2^1 + 1$) levels of gray, $\alpha = 1$, $\eta_r = 1$.

Figure 6.9: Rectangular Ordered Dither of a Gray Scale Ramp.
(b) 5 (or $2^2 + 1$) levels of gray, $\alpha = 1$, $\eta_r = 2$.

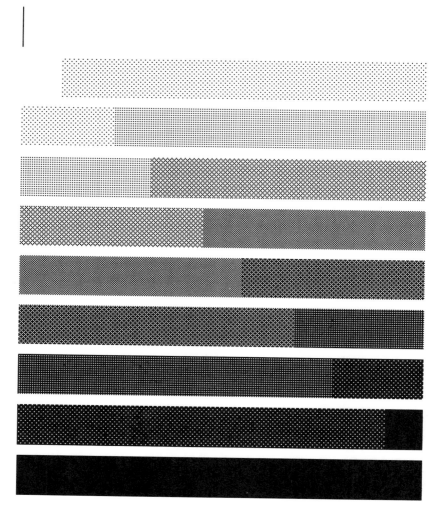

Figure 6.9: Rectangular Ordered Dither of a Gray Scale Ramp.
(c) 9 (or $2^3 + 1$) levels of gray, $\alpha = 1$, $\eta_r = 3$.

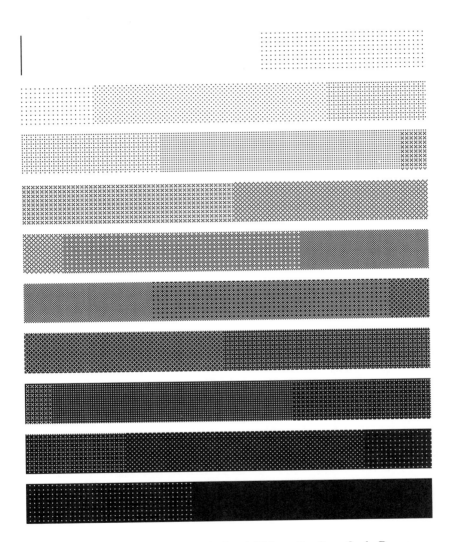

Figure 6.9: Rectangular Ordered Dither of a Gray Scale Ramp.
(d) 17 (or $2^4 + 1$) levels of gray, $\alpha = 1$, $\eta_r = 4$.

Figure 6.9: Rectangular Ordered Dither of a Gray Scale Ramp.
(e) 33 (or $2^5 + 1$) levels of gray, $\alpha = 1$, $\eta_r = 5$.

Figure 6.9: Rectangular Ordered Dither of a Gray Scale Ramp.
(f) 65 (or $2^6 + 1$) levels of gray, $\alpha = 1$, $\eta_r = 6$.

Figure 6.9:   Rectangular Ordered Dither of a Gray Scale Ramp.
(g) 129 (or $2^7 + 1$) levels of gray, $\alpha = 1$, $\eta_r = 7$.

Figure 6.9: Rectangular Ordered Dither of a Gray Scale Ramp. (h) 257 (or $2^8 + 1$) levels of gray, $\alpha = 1$, $\eta_r = 8$.

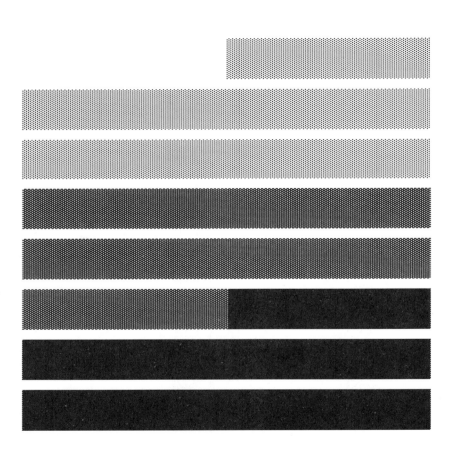

Figure 6.10: Hexagonal Ordered Dither of a Gray Scale Ramp.
(a) 4 (or $3^1 + 1$) levels of gray, $\alpha = \frac{2}{\sqrt{3}}$, $\eta_h = 1$.

Figure 6.10: Hexagonal Ordered Dither of a Gray Scale Ramp.
(b) 10 (or $3^2 + 1$) levels of gray, $\alpha = \frac{2}{\sqrt{3}}$, $\eta_h = 2$.

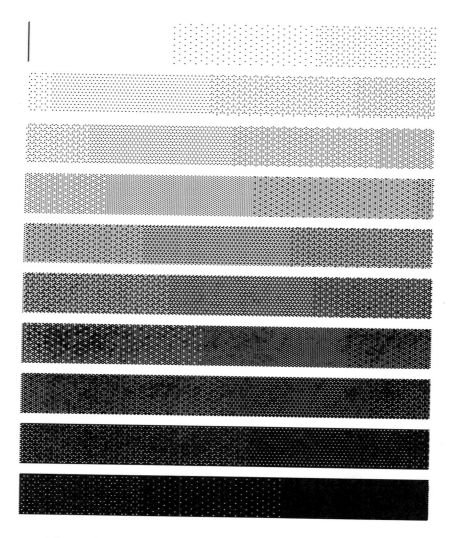

Figure 6.10:   Hexagonal Ordered Dither of a Gray Scale Ramp.
(c) 28 (or $3^3 + 1$) levels of gray, $\alpha = \frac{2}{\sqrt{3}}$, $\eta_h = 3$.

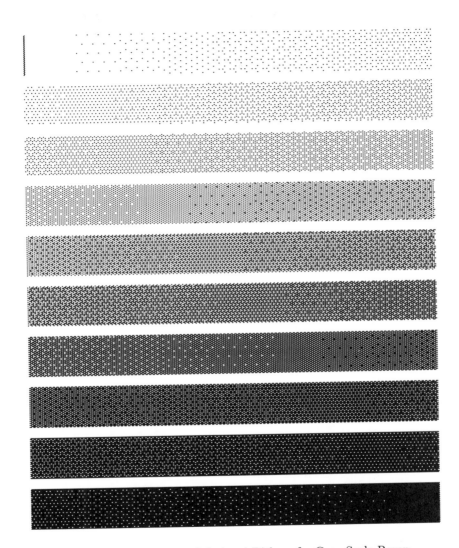

Figure 6.10: Hexagonal Ordered Dither of a Gray Scale Ramp. (d) 82 (or $3^4 + 1$) levels of gray, $\alpha = \frac{2}{\sqrt{3}}$, $\eta_h = 4$.

Figure 6.10:   Hexagonal Ordered Dither of a Gray Scale Ramp.
(d) 244 (or $3^5 + 1$) levels of gray, $\alpha = \frac{2}{\sqrt{3}}$, $\eta_h = 5$.

As was mentioned at the onset of this section, these results can be applied to the two other regular grids, hexagonal grids with $\alpha = 2\sqrt{3}$ and $\alpha = 2$. While rotating the rectangular results by 45° for application on a hexagonal grid with $\alpha = 2$ is straightforward, it may not be clear why the hexagonal results in this section produced on a hexagonal grid of the first kind need to be rotated 90° for application on a hexagonal grid of the second kind ($\alpha = 2\sqrt{3}$). Figure 6.14 (page 169) illustrates why this rotation is necessary.

It should be pointed out that while the idea of applying ordered dither to hexagonal grids is new, using a homogeneous distribution of dots to create the illusion of gray scale is not a modern one. Figure 6.15 shows the detail of a binary image produced in 1844. It was a 125 thread per inch silk weaving generated on the punchcard operated Jacquard loom [38]. It shows a close resemblance to the patterns produced by rectangular ordered dither.

Figure 6.11: Rectangular Ordered Dither of a Scanned Picture.
(a) 3 (or $2^1 + 1$) levels of gray, $\alpha = 1$, $\eta_r = 1$.

Figure 6.11: Rectangular Ordered Dither of a Scanned Picture.
(b) 5 (or $2^2 + 1$) levels of gray, $\alpha = 1$, $\eta_r = 2$.

Figure 6.11:   Rectangular Ordered Dither of a Scanned Picture.
(c) 9 (or $2^3 + 1$) levels of gray, $\alpha = 1$, $\eta_r = 3$.

Figure 6.11: Rectangular Ordered Dither of a Scanned Picture.
(d) 17 (or $2^4 + 1$) levels of gray, $\alpha = 1$, $\eta_r = 4$.

Figure 6.11:   Rectangular Ordered Dither of a Scanned Picture.
(e) 129 (or $2^7 + 1$) levels of gray, $\alpha = 1$, $\eta_r = 7$.

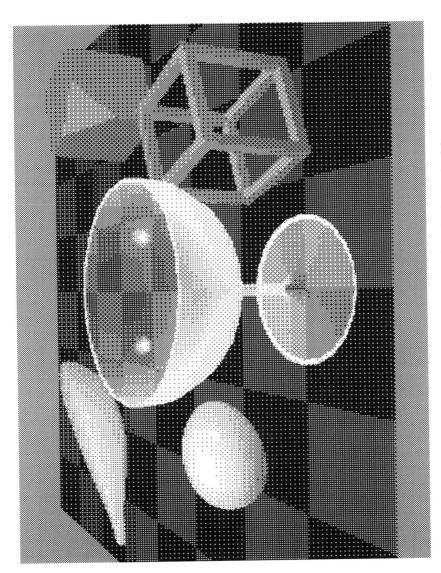

Figure 6.12: Rectangular Ordered Dither of a Synthesized Image. (a) 17 (or $2^4 + 1$) levels of gray, $\alpha = 1$, $\eta_r = 4$.

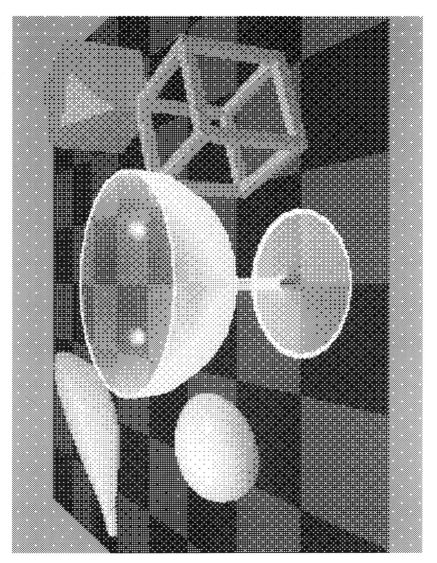

Figure 6.12: Rectangular Ordered Dither of a Synthesized Image. (b) 129 (or $2^7 + 1$) levels of gray, $\alpha = 1$, $\eta_r = 7$.

Figure 6.13: Hexagonal Ordered Dither of a Scanned Picture.
(a) 4 (or $3^1 + 1$) levels of gray, $\alpha = \frac{2}{\sqrt{3}}$, $\eta_h = 1$.

Figure 6.13: Hexagonal Ordered Dither of a Scanned Picture. (b) 10 (or $3^2 + 1$) levels of gray, $\alpha = \frac{2}{\sqrt{3}}$, $\eta_h = 2$.

Figure 6.13: Hexagonal Ordered Dither of a Scanned Picture. (c) 28 (or $3^3 + 1$) levels of gray, $\alpha = \frac{2}{\sqrt{3}}$, $\eta_h = 3$.

Figure 6.13:   Hexagonal Ordered Dither of a Scanned Picture.
(e) 244 (or $3^5 + 1$) levels of gray, $\alpha = \frac{2}{\sqrt{3}}$, $\eta_h = 5$.

```
              3
      2   5   8   2
        7   1   4
      3   6   9   3
              2
```

Second order threshold array
for a regular hexagonal grid of the First Kind $\left(\alpha = \frac{2}{\sqrt{3}}\right)$.

```
              3
    2     5       8     2
       7     1       4
    3     6       9     3
              2
```

Incorrect use of the above data
on a regular hexagonal grid of the Second Kind $\left(\alpha = 2\sqrt{3}\right)$.

```
         2       3
             4
         8       9
    3    1       2
         5       6
             7
         2       3
```

Correct translation of threshold values
for a hexagonal grid of the Second Kind.

Figure 6.14: Importance of Hexagonal Kind.

Figure 6.15: Detail from a 1844 Silk Weaving by the Jacquard Loom.
(Smithsonian Institution Photo [73].)

6.3 Exposure Plots

Figure 6.16, pages 172 through 179, displays exposure plots of $C_\Sigma(\mathbf{f})$ for $\eta_r = 1$ through 8 on a rectangular spatial grid ($\alpha = 1$), along with the corresponding values of $I_\Sigma[\mathbf{k}]$ for the first quadrant ($C_\Sigma(\mathbf{f})$ possesses 4 fold symmetry). At the bottom of each exposure plot, a scale defining the actual dimensions (in cycles/unit-length) of the plot is terms of the original sample period, S, is provided.

Similarly, Figure 6.17, pages 180 through 184, displays exposure plots for $\eta_h = 1$ to 5 where the sample grid was a regular hexagonal grid of the first kind ($\alpha = \frac{2}{\sqrt{3}}$). The numerical values of $I_\Sigma[\mathbf{k}]$ for the first sextant are shown at the top of each plot completely specifying the composite Fourier transform since it has 6 fold symmetry.

Consistent with the fair comparison approach of section 6.2, the number of spatial samples per unit-area are held constant, and the exposure plots are all generated with the same dimensional scale. Note that rectangular and hexagonal images with the same spatial pixel area have basebands of equal area.

The rectangular and hexagonal sample periods, S_r and S_h, used to define the scales in each exposure plot are constrained by the equal-pixel-area condition, $S_r L_r = S_h L_h$, and the aspect ratios used, $\alpha_r = 1$ and $\alpha_h = \frac{2}{\sqrt{3}}$. As was indicated in section 2.1.2.2 they are thus related as

$$\frac{S_r}{S_h} = \sqrt{\frac{\sqrt{3}}{2}}.$$

6.3.1 Analysis

The exposure plots of figures 6.16 and 6.17 beautifully illustrate the results of the Method of Recursive Tessellation.

For $\eta = 1$, the only location of energy in the frequency domain besides the zero frequency term is at the "corners" of the baseband. This corresponds to the highest frequency which the grid is capable of accommodating; look ahead to Figure 8.2, page 237, for an illustration of this. For the rectangular grid, it is the checkerboard pattern. Hexagonal grids have two "highest frequency" patterns which are negatives of one another. The patterns occur when either 1 or 2 of the 3 elements of a first order threshold array are "on".

0.33

1.00

Figure 6.16: Composite Fourier Transform of Rectangular Ordered Dither.
(a) Average of 3 patterns, $\eta_r = 1$.
(First quadrant of $I_\Sigma[\mathbf{k}]$ shown at top.)

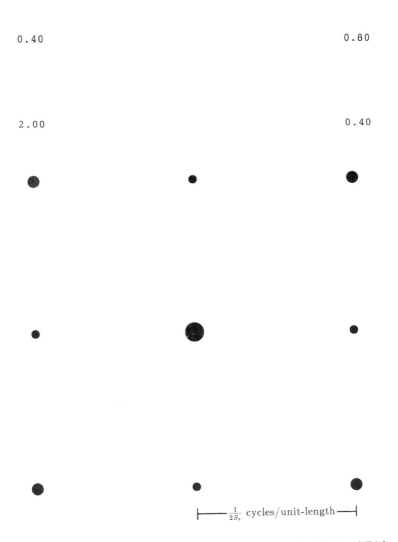

0.40

0.80

2.00

0.40

$\longmapsto \frac{1}{2S_r}$ cycles/unit-length $\longmapsto$

Figure 6.16: Composite Fourier Transform of Rectangular Ordered Dither.
(b) Average of 5 patterns, $\eta_r = 2$.
(First quadrant of $I_\Sigma[\mathbf{k}]$ shown at top.)

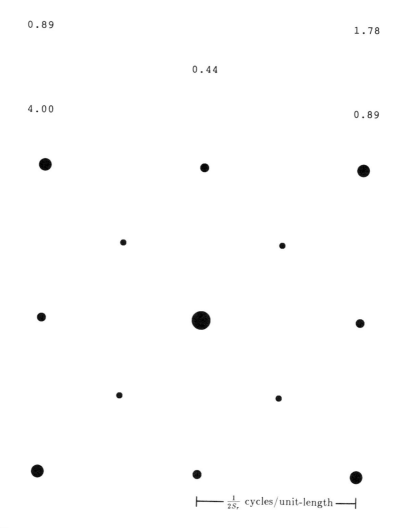

0.89

1.78

0.44

4.00

0.89

$\vdash$ $\frac{1}{2S_r}$ cycles/unit-length $\dashv$

Figure 6.16:   Composite Fourier Transform of Rectangular Ordered Dither.
(c) Average of 9 patterns, $\eta_r = 3$.
(First quadrant of $I_\Sigma[\mathbf{k}]$ shown at top.)

1.88   0.47   3.76

0.47   0.94   0.47

8.00   0.47   1.88

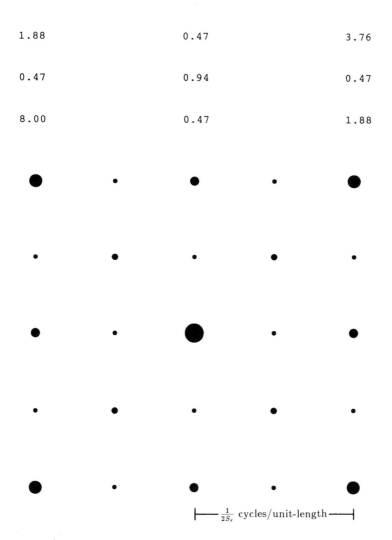

$\longmapsto \frac{1}{2S_r}$ cycles/unit-length $\longmapsto$

Figure 6.16: Composite Fourier Transform of Rectangular Ordered Dither.
(d) Average of 17 patterns, $\eta_r = 4$.
(First quadrant of $I_\Sigma[\mathbf{k}]$ shown at top.)

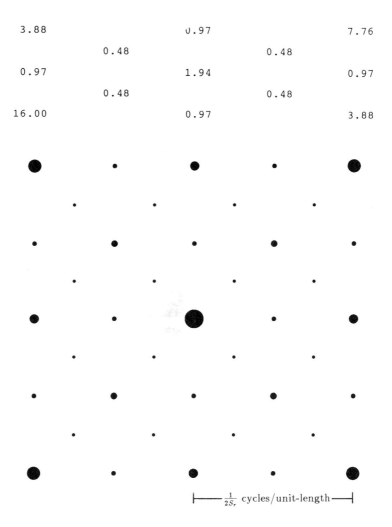

3.88		ʊ.97		7.76
	0.48		0.48	
0.97		1.94		0.97
	0.48		0.48	
16.00		0.97		3.88

$\longmapsto \frac{1}{2S_r}$ cycles/unit-length $\longmapsto$

Figure 6.16:   Composite Fourier Transform of Rectangular Ordered Dither.
(e) Average of 33 patterns, $\eta_r = 5$.
(First quadrant of $I_\Sigma[\mathbf{k}]$ shown at top.)

7.88	0.49	1.97	0.49	15.75
0.49	0.98	0.49	0.98	0.49
1.97	0.49	3.94	0.49	1.97
0.49	0.98	0.49	0.98	0.49
32.00	0.49	1.97	0.49	7.88

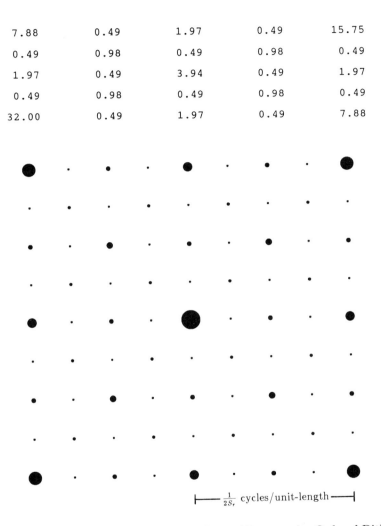

$\longmapsto \frac{1}{2S_r}$ cycles/unit-length $\longmapsto$

Figure 6.16: Composite Fourier Transform of Rectangular Ordered Dither.
(f) Average of 65 patterns, $\eta_r = 6$.
(First quadrant of $I_\Sigma[\mathbf{k}]$ shown at top.)

```
15.88          0.99          3.97          0.99          31.75
         0.50          0.50          0.50          0.50
 0.99          1.98          0.99          1.98          0.99
         0.50          0.50          0.50          0.50
 3.97          0.99          7.94          0.99          3.97
         0.50          0.50          0.50          0.50
 0.99          1.98          0.99          1.98          0.99
         0.50          0.50          0.50          0.50
64.00          0.99          3.97          0.99          15.88
```

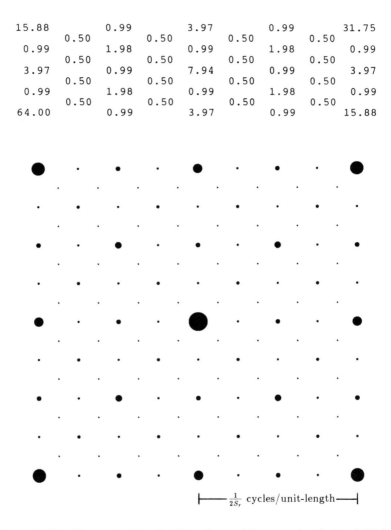

$\longmapsto \frac{1}{2S_r}$ cycles/unit-length $\longmapsto$

Figure 6.16:   Composite Fourier Transform of Rectangular Ordered Dither.
(g) Average of 129 patterns, $\eta_r = 7$.
(First quadrant of $I_\Sigma[\mathbf{k}]$ shown at top.)

31.88	0.50	1.99	0.50	7.97	0.50	1.99	0.50	63.75
0.50	1.00	0.50	1.00	0.50	1.00	0.50	1.00	0.50
1.99	0.50	3.98	0.50	1.99	0.50	3.98	0.50	1.99
0.50	1.00	0.50	1.00	0.50	1.00	0.50	1.00	0.50
7.97	0.50	1.99	0.50	15.94	0.50	1.99	0.50	7.97
0.50	1.00	0.50	1.00	0.50	1.00	0.50	1.00	0.50
1.99	0.50	3.98	0.50	1.99	0.50	3.98	0.50	1.99
0.50	1.00	0.50	1.00	0.50	1.00	0.50	1.00	0.50
128.00	0.50	1.99	0.50	7.97	0.50	1.99	0.50	31.88

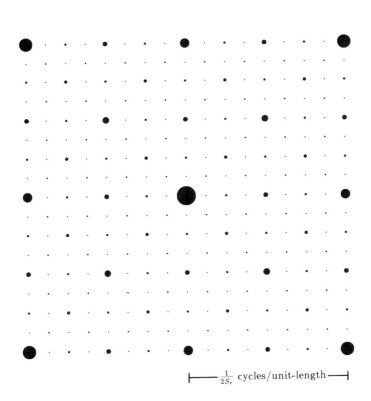

$\vdash$——$\frac{1}{2S_r}$ cycles/unit-length——$\dashv$

Figure 6.16: Composite Fourier Transform of Rectangular Ordered Dither.
(h) Average of 257 patterns, $\eta_r = 8$.
(First quadrant of $I_\Sigma[\mathbf{k}]$ shown at top.)

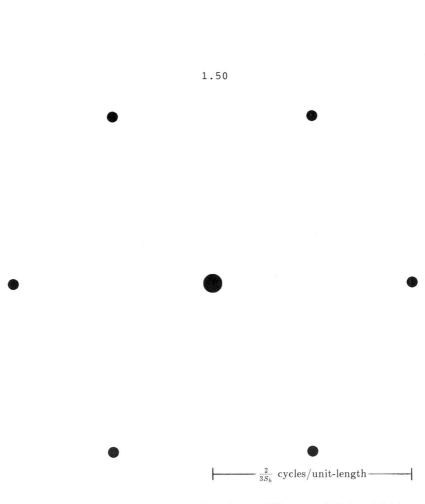

Figure 6.17: Composite Fourier Transform of Hexagonal Ordered Dither.
(a) Average of 4 patterns, $\eta_h = 1$.
(First sextant of $I_\Sigma[\mathbf{k}]$ shown at top.)

1.73   1.73

0.60

4.50

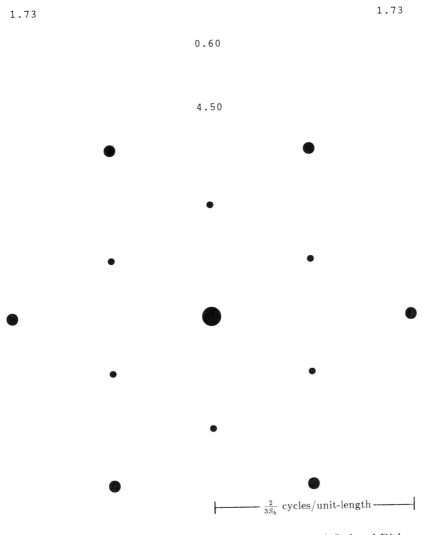

$\longmapsto \frac{2}{3S_h}$ cycles/unit-length $\longmapsto$

Figure 6.17: Composite Fourier Transform of Hexagonal Ordered Dither.
(b) Average of 10 patterns, $\eta_h = 2$.
(First sextant of $I_\Sigma[\mathbf{k}]$ shown at top.)

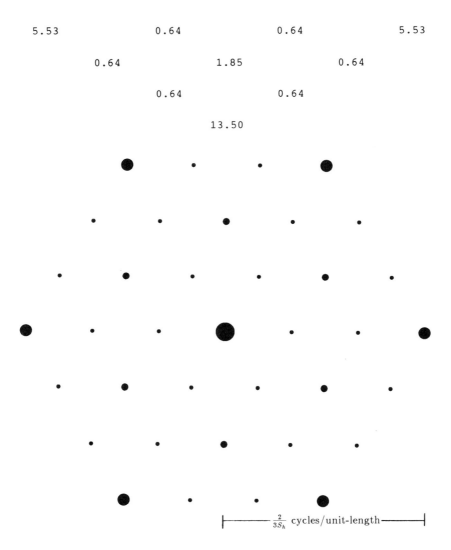

5.53   0.64   0.64   5.53

    0.64   1.85   0.64

        0.64   0.64

            13.50

$\longmapsto \frac{2}{3S_h}$ cycles/unit-length $\longmapsto$

Figure 6.17:   Composite Fourier Transform of Hexagonal Ordered Dither.
(c) Average of 28 patterns, $\eta_h = 3$.
(First sextant of $I_\Sigma[\mathbf{k}]$ shown at top.)

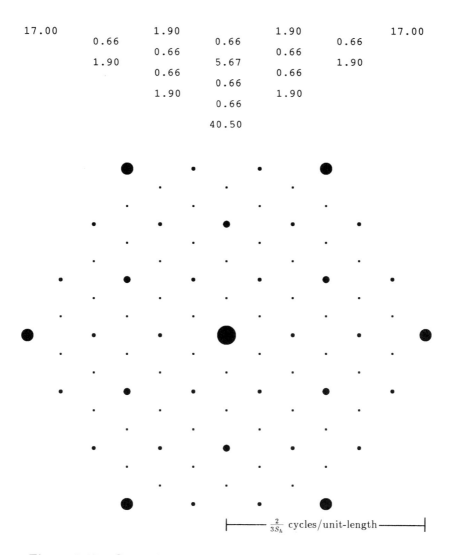

```
17.00              1.90            1.90              17.00
       0.66              0.66            0.66
              0.66              0.66
       1.90              5.67            1.90
              0.66              0.66
       1.90              0.66            1.90
                     0.66
                    40.50
```

$\dfrac{2}{3S_h}$ cycles/unit-length

Figure 6.17: Composite Fourier Transform of Hexagonal Ordered Dither.
(d) Average of 82 patterns, $\eta_h = 4$.
(First sextant of $I_\Sigma[\mathbf{k}]$ shown at top.)

51.41   0.66   0.66   5.72   0.66   0.66   5.72   0.66   0.66 51.41
    0.66   1.91   0.66   0.66   1.91   0.66   0.66   1.91   0.66
       0.66   0.66   1.91   0.66   0.66   1.91   0.66   0.66
          5.72   0.66   0.66 17.14   0.66   0.66   5.72
             0.66   1.91   0.66   0.66   1.91   0.66
                0.66   0.66   1.91   0.66   0.66
                   5.72   0.66   0.66   5.72
                      0.66   1.91   0.66
                         0.66   0.66
                         121.50

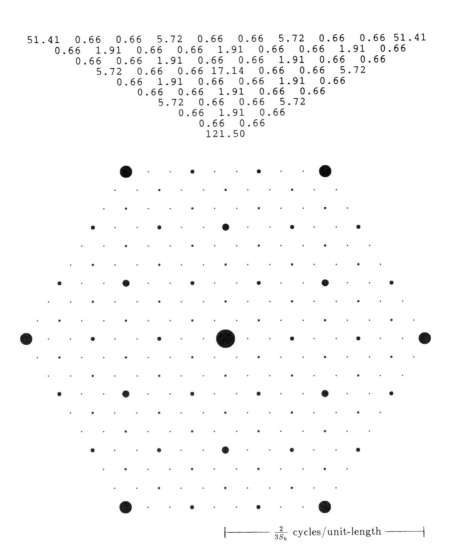

$\longmapsto$ $\frac{2}{3S_h}$ cycles/unit-length $\longmapsto$

Figure 6.17: Composite Fourier Transform of Hexagonal Ordered Dither.
(e) Average of 244 patterns, $\eta_h = 5$.
(First sextant of $I_\Sigma[\mathbf{k}]$ shown at top.)

For every order, η, these high frequency corners are second only to the zero frequency term in magnitude. This is exactly what one would expect from a good dispersed-dot ordered dither algorithm.

As η, and thus the number of elements in the threshold period, increases, the additional frequency terms in the Fourier transform assume positions in between those of smaller η, with proportionally less energy associated with them. The size of the baseband, however, always remains fixed.

Observe that families of frequency coefficients of the same magnitude are arranged exactly as the families of threshold points within the stages of the Method of Recursive Tessellation! Families of larger dots correspond to earlier stages.

Also, the gray-level-number vs. low-frequency-texture tradeoff demonstrated in the examples of section 6.2 can be explained in the frequency domain. As the number of gray levels increases, so does the number of coefficients. As coefficients get closer to the zero frequency location, textures of lower frequency appear in the image.

On high spatial resolution displays, it is the set of frequency terms closest to the zero frequency center that determine whether or not any patterns will be perceived. Redrawing the data given by Taylor [81], approximated in Figure 5.3 (page 84), in polar form would reveal a somewhat square shaped frequency threshold plot with the cusps oriented along the horizontal and vertical axes. Any nonzero frequency components inside this perceptual mask would be seen as a periodic pattern; components outside this area would be spatially integrated to yield a sensation of a solid average gray value.

So, in observing the exposure plots for rectangular grids (Figure 6.16) for any even ordered period, η, the *same* susceptibility of the human visual system to perceive low frequency periods exists for the next higher odd ordered period, $\eta + 1$, which can display twice as many gray levels! Therefore, from this examination in the frequency domain it can be argued that only odd rectangular ordered dither arrays should used.

The superior radial symmetry of hexagonal grids is also evident in the frequency domain. Energy is distributed more isotropically. While the corners of the rectangular Fourier transforms are slightly further from the zero frequency term than the corners of the hexagonal Fourier transforms, a fact of great importance in Chapter 8, the hexagonal transforms have higher frequency components along both the horizontal and vertical directions, where high frequencies are most needed due to the visual system's acute sensitivity there. The superposition of a square and hexagon of equal area in Figure 6.18 demonstrates these geometric characteristics.

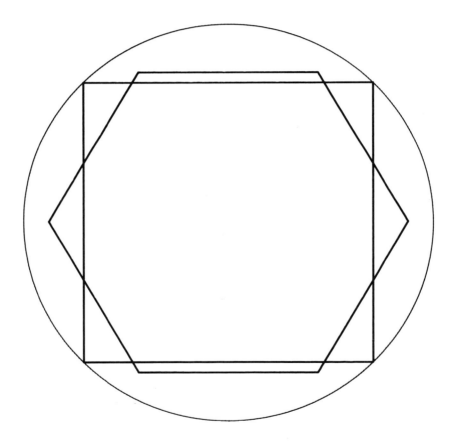

Figure 6.18: Superposition of Square and Regular Hexagonal Basebands of equal area corresponding to equal sample densities. Hexagonal grids support higher horizontal and vertical frequencies, but the highest frequencies are accommodated on rectangular grids.

Chapter 7

Ordered Dither on Asymmetric Grids

The general solution for generating ordered dither threshold arrays for regular grids, both rectangular and hexagonal, is provided by the method of recursive tessellation as demonstrated in the last chapter. Figure 7.1 illustrates the failure of such threshold arrays to perform well on an asymmetric grid. In this example, both the scanned and synthesized image have been scaled to fit on a rectangular grid with aspect ratio, $\alpha_r = \frac{1}{6}$. The development of compensated threshold arrays for halftoning by ordered dither on asymmetric grids of *any* aspect ratio is the subject of this chapter.

The physical mechanisms and constraints which fix the sample period, S, and line period, L, in real display devices are usually very different. It may be the case that it is easy to increase the resolution in one direction and difficult or impossible to do so in the other. Some theoretical questions come to mind. If a resolution increase is available in only one direction, should that increase be used at the expense of grid symmetry? And if so, to what extent should such a unidirectional increase be used?

As one might expect, a resolution increase, symmetric or otherwise, will always enhance the display of digital imagery. Figure 7.2 shows

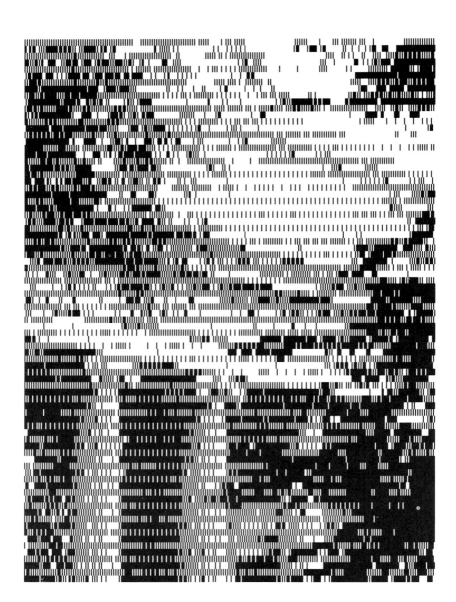

Figure 7.1: Failure of uncompensated ordered dither
on an asymmetric grid.
(a) Scanned Picture, $\eta_r = 5, \alpha_r = \frac{1}{6}$ (vertical resolution decrease).

Figure 7.1: Failure of uncompensated ordered dither on an asymmetric grid.

(b) Synthesized Image, $\eta_r = 5$, $\alpha_r = \frac{1}{6}$ (vertical resolution decrease).

the scanned image of Figure 7.1(a) halftoned in the same way on the same grid with $\alpha_r = \frac{1}{6}$, but with the input subsampled and the binary output replicated six times; symmetry was achieved at the expense of resolution.

In some cases, a unidirectional increase in resolution can enhance the quality of a display in the same way as a symmetric increase with an equal number of samples per unit area. As early as 1940, it was found that in the case of m-ary images, altering the aspect ratio within the range $.4 < \alpha < 2.5$ had little effect on perceived sharpness [8].

In this chapter, the solution to providing symmetrically ordered dither patterns for any unidirectional resolution increase will be given, with examples on rectangular grids of various resolutions. As will be seen, the solution for rectangular grids hinges on solving the asymmetry problem for hexagonal grids. For this reason, the hexagonal solution is presented first.

As discussed in section 1.3, the resolution (in pixels per unit area) with which images on asymmetric grids are shown is lower than that for symmetric grids by an amount proportional to the degree of asymmetry.

7.1   Hexagonal Grid Solution

The trick to be employed here is based on applying the threshold arrays generated in Chapter 6 to symmetric subsamples of the asymmetric grid.

In generating the hexagonal threshold arrays by the method of recursive tessellation (Figure 6.4, page 133), the centers of the even and odd period tiles are on "super grids" of only two aspect ratios. One is the aspect ratio of the grid itself, α_h, or $\frac{2}{\sqrt{3}}$, and the other is $3\alpha_h$, or $2\sqrt{3}$, that of a regular hexagonal grid of the second kind. These two super grid shapes can be seen in several of the patterns in Figure 6.10, particularly on page 154.

The two super grid aspect ratios of α_h and $3\alpha_h$ will be observed when these patterns appear on any hexagonal grid of the first kind ($\alpha_h < 2$). For hexagonal ordered dither patterns on grids of the second kind ($\alpha_h > 2$), the two super grid aspect ratios will be α_h and $\alpha_h/3$.

Recalling that Covering Efficiency introduced in Section 2.1.2.1 was

Figure 7.2: Sacrificing Resolution for Symmetry with $\alpha_r = \frac{1}{6}$.
Ordered Dither on a Scanned Image with $\eta_r = 5$.

defined as the ratio of pixel shape to a circumscribing circle and plotted
in Figure 2.5, page 25, such a metric also describes the radial symmetry
of any grid as a function of aspect ratio. A close up of Figure 2.5 for
hexagonal grids is shown in Figure 7.3. On this curve, complement pairs
of super grid aspect ratios are shown. The range $\frac{2}{3} < \alpha_h \le 2$ is defined
as the *principal aspect ratio range* for hexagonal grids of the first kind,
and $2 < \alpha_h \le 6$ as the principal aspect ratio range for hexagonal grids
of the second kind.

All aspect ratios in the principal range of a hexagonal grid on one
kind have a complement aspect ratio in the principal range of a hexagonal grid of the other kind. For grids in that range, no compensation is
necessary; the symmetry metric, $E(\alpha_h)$, is very high.

When the aspect ratio is in the range $\frac{2}{9} < \alpha_h \le \frac{2}{3}$, a compensated
threshold array is generated by a method of one-dimensional ternary
subsampling and replication which effectively multiplies the aspect ratios of the two super grids by 3, thus getting them into the principal
range. By *ternary subsampling* is meant subsampling an asymmetric
hexagonal grid by 3 in the closer-packed direction in the manner shown
in Figure 7.4. Note that the subsampled grid, with an aspect ratio of
$3\alpha_h$, is much more symmetric than the original grid.

Generation of the compensated threshold array begins by assigning
the values of a regular hexagonal ordered dither array of any order, η_h, to
these samples. *ternary replication* is the method by which the in between
samples are assigned. Very similar to the technique of assigning values
described in section 6.1.2, the offset vector in this case always points
along the direction of subsampling.

An example of this is shown in Figure 7.5. A hexagonal threshold
array of order $\eta_h = 1$ shown in (a) is valid for any grid in the principal
range, $\frac{2}{3} < \alpha_h \le 2$. In (b), an array compensated with one ternary
replication is valid for grids with an aspect ratio in the range $\frac{2}{9} < \alpha_h \le
\frac{2}{3}$. The two passes in this example add 3 to previously assigned threshold
values and assign the result to the location at the immediate right.

As one might expect, this procedure can be repeated for grids of the
even smaller aspect ratio, $\frac{2}{27} < \alpha_h \le \frac{2}{9}$. Figure 7.5(c) shows such a
threshold array. In each of these cases, the values of the original regular
threshold array are circled. A generalization of this method for any
aspect ratio will now be formally stated.

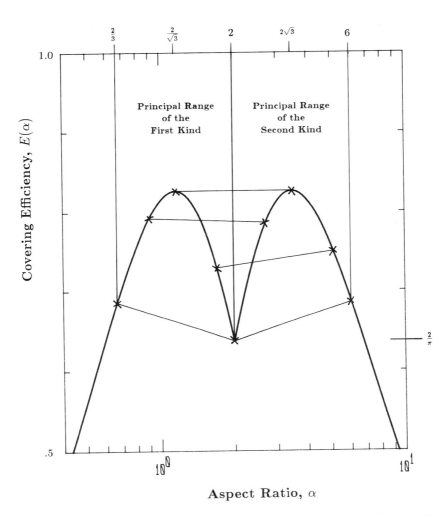

Figure 7.3: Complement Aspect Ratio Pairs found in Hexagonal Ordered Dither Patterns on grids with aspect ratios in the principal range. $E(\alpha)$ is a measure of symmetry.

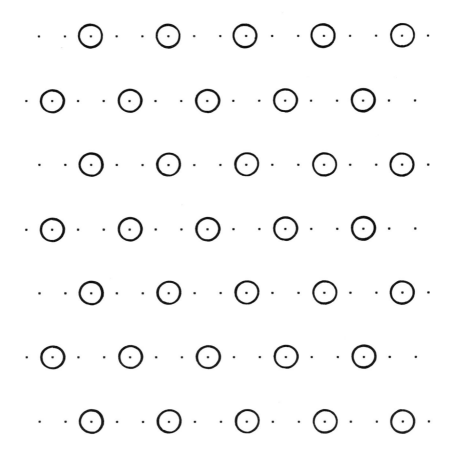

Figure 7.4: Ternary Subsampling of an Asymmetric Hexagonal Grid. ($\kappa = 1$ in this case.)

(a) $\eta_h = 1$, $\kappa = 0$, for the range $\frac{2}{3} < \alpha_h \leq 2$.

(b) $\eta_h = 1$, $\kappa = 1$, for the range $2(\frac{1}{3})^2 < \alpha_h \leq 2(\frac{1}{3})$.

②	11	20	5̲	14	23	8̲	17	26	③
③	12	21	6̲	15	24	9̲	18	27	①	10	19	4̲	13	22	7̲	16	25	②
②	11	20	5̲	14	23	8̲	17	26	③

(c) $\eta_h = 1$, $\kappa = 2$, for the range $2(\frac{1}{3})^3 < \alpha_h \leq 2(\frac{1}{3})^2$.

Figure 7.5: Compensating Threshold Arrays by Ternary Replication.

7.1.1 Generalized Procedure

In general, for asymmetric hexagonal grids, a new hexagonal grid is created by repeated ternary subsampling in the closer-packed direction until the resulting grid has an aspect ratio in the principal range. In this way, the pair of complementary aspect ratios have the highest possible "symmetry values", $E(\alpha_h)$, (Figure 7.3).

The rule determining the *number of ternary subdivisions and replications*, κ, is as follows. For asymmetric hexagonal grids of the first kind with an aspect ratio in the range

$$2\left(\frac{1}{3}\right)^{\kappa+1} < \alpha_h \leq 2\left(\frac{1}{3}\right)^{\kappa} \tag{7.1}$$

a subsampling factor of $(3)^{\kappa}$ is required in the S direction, and for asymmetric hexagonal grids of the second kind with an aspect ratio in the range

$$2(3)^{\kappa} < \alpha_h \leq 2(3)^{\kappa+1} \tag{7.2}$$

a subsampling factor of $(3)^{\kappa}$ is required on the L direction.

The procedure for assigning values to the array begins by assigning the subsamples the values of a threshold array of any order, η_h. Note that for grids with aspect ratios in the principal range, $\kappa = 0$ and no subsampling is required; the ordered dither threshold arrays can be used without further compensation.

For $\kappa \neq 0$, the in between samples are assigned with κ ternary replications. The offset vector (see section 6.1.2) is oriented along the closer-packed direction with length equal to $(3)^{\kappa-i}$ units, for $i = 1$ to κ. At each step, i, the grid point pointed to by the offset vector is assigned a value equal to that at the tail of the vector plus $(3)^{\eta_h+i-1}$. The resulting number of elements in the threshold array is $Z = (3)^{\eta_h+\kappa}$.

7.2 Rectangular Grid Solution

The aspect ratio immunity argument (section 2.1.2.1) in favor of hexagonal grids, which showed that hexagonal grids were more symmetric over an order of magnitude in aspect ratio than a perfectly square rectangular grid, can be exploited in the rectangular case by selecting every other grid point in a hexagonal fashion. Besides enjoying the attributes of hexagonal grids, dividing a rectangular grid in this way preserves the very high frequency checkerboard pattern at middle gray, a most important feature of rectangular grids.

Remarkably then, the solution for generating compensated ordered dither threshold arrays for asymmetric rectangular grids is a simple extrapolation of the hexagonal solution. Figure 7.6 shows how an asymmetric rectangular grid with aspect ratio α_r is subsampled to form a hexagonal grid of aspect ratio $\alpha_h = \alpha_r/2$. The values of the rectangular threshold array are determined by first assigning to the hexagonal subsamples a corrected hexagonal threshold array of order, η_h, with κ appropriate for $\alpha_h = \alpha_r/2$. The in between samples are given the value of the neighbor in the tighter packed direction plus $(3)^{\eta_h+\kappa}$. The resulting compensated rectangular ordered dither threshold array has $Z = 2(3)^{\eta_h+\kappa}$ elements.

In terms of the asymmetric aspect ratio of the rectangular grid, the value of κ is determined as that which satisfies

$$(\tfrac{1}{3})^{\kappa+1} < \alpha_r \le (\tfrac{1}{3})^{\kappa} \qquad \text{for } \alpha_r \le 1$$
with subsampling along the S direction

$$(7.3)$$

$$\text{or} \quad (3)^{\kappa} < \alpha_r \le (3)^{\kappa+1} \qquad \text{for } \alpha_r > 1$$
with subsampling along the L direction.

$$(7.4)$$

The examples in Figure 7.7 were generated from the compensated hexagonal threshold arrays in Figure 7.5 for the three aspect ratios shown. The circled values in Figure 7.7 highlight those hexagonal arrays used.

All of the example images in the following sections were generated on rectangular grids with small aspect ratios. Examples of grids with large aspect ratios are not shown since the effects are the same through transposition.

7.3 Examples for $\alpha_r = \frac{1}{2}$

The proper value of κ for the range into which the aspect ratio $\alpha_r = \frac{1}{2}$ falls is 0. The compensated threshold array to be used is shown

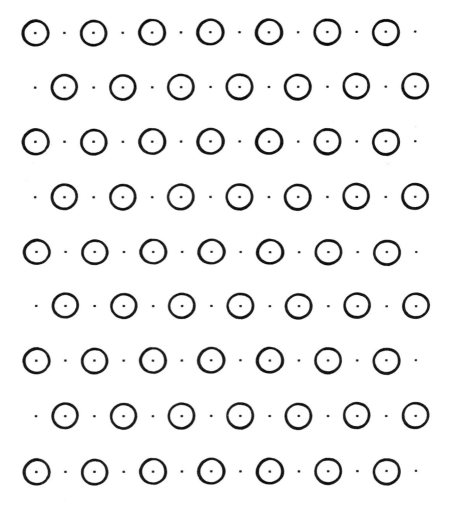

Figure 7.6: Hexagonal Subsampling of an Asymmetric Rectangular Grid. ($\kappa = 0$ in this case.)

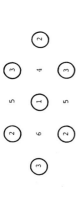

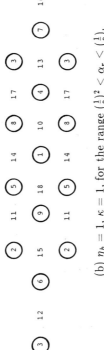

(a) $\eta_h = 1$, $\kappa = 0$, for the range $\frac{1}{3} < \alpha_r \leq 1$.

(b) $\eta_h = 1$, $\kappa = 1$, for the range $\left(\frac{1}{3}\right)^2 < \alpha_r \leq \left(\frac{1}{3}\right)$.

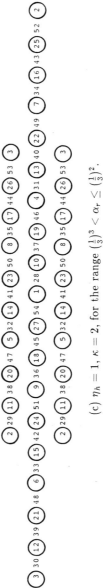

(c) $\eta_h = 1$, $\kappa = 2$, for the range $\left(\frac{1}{3}\right)^3 < \alpha_r \leq \left(\frac{1}{3}\right)^2$.

Figure 7.7: Compensating Rectangular Arrays by Hexagonal Subsampling. Circled values correspond to threshold values from Figure 7.5, page 195.

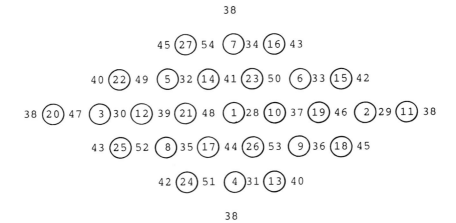

Figure 7.8: Compensated Threshold Array for $\frac{1}{3} < \alpha_r \le 1$.
$\eta_h = 3$, $\kappa = 0$, $Z = 54$.
Circled values correspond to the original threshold array.

in Figure 7.8 in the usual odd rectangular period form. The hexagonally subsampled locations are circled and correspond to a hexagonal threshold array with order $\eta_h = 3$. Notice how these circled values comprise an almost regular hexagonal grid. The total number of elements is $Z = 2(3)^{3+0} = 54$.

To examine the effect of this compensated threshold array, the gray scale ramp (Figure 7.9) and scanned picture (Figure 7.10) compare the results of halftoning with

(a) an uncompensated rectangular ordered dither threshold array of order $\eta_r = 5$ and $Z = 32$,

(b) the same as above but first replicating pixels in the S direction, sacrificing resolution for symmetry, and

(c) the compensated threshold array of Figure 7.8.

The horizontal banding evident in (a) is somewhat eliminated at the great expense of forfeiting horizontal resolution in (b). The images in (c) use that extra resolution dexterously to form radially symmetric patterns.

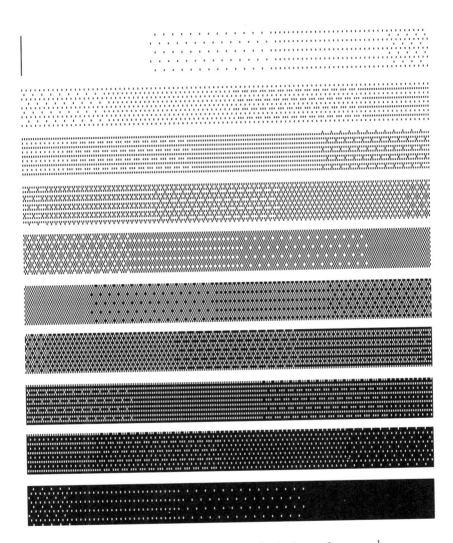

Figure 7.9: Comparison of Gray Scale Ramp for $\alpha_r = \frac{1}{2}$.
(a) Uncompensated Rectangular Ordered Dither, $\eta_r = 5$.

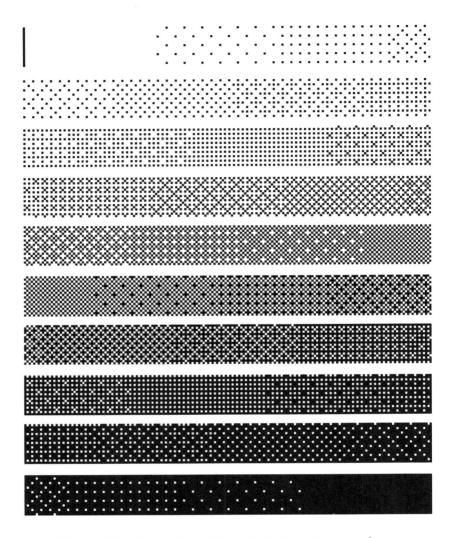

Figure 7.9: Comparison of Gray Scale Ramp for $\alpha_r = \frac{1}{2}$.
(b) $\eta_r = 5$, sacrificing resolution for symmetry.

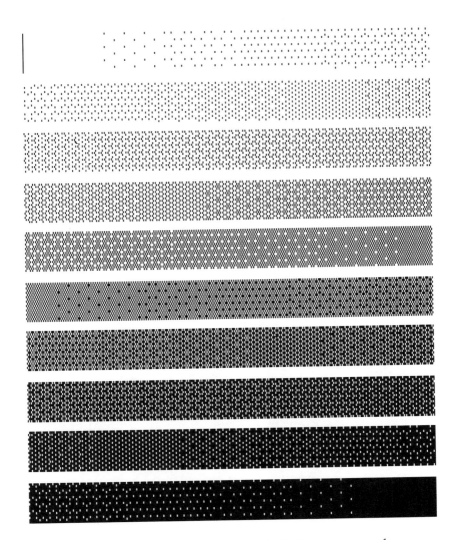

Figure 7.9: Comparison of Gray Scale Ramp for $\alpha_r = \frac{1}{2}$.
(c) Compensated Threshold Array, $\eta_h = 3$, $\kappa = 0$.

Figure 7.10: Comparison of Scanned Picture for $\alpha_r = \frac{1}{2}$.
(a) Uncompensated Rectangular Ordered Dither, $\eta_r = 5$.

Figure 7.10: Comparison of Scanned Picture for $\alpha_r = \frac{1}{2}$.
(b) $\eta_r = 5$, sacrificing resolution for symmetry.

Figure 7.10: Comparison of Scanned Picture for $\alpha_r = \frac{1}{2}$.
(c) Compensated Threshold Array, $\eta_h = 3$, $\kappa = 0$.

Considerable insight is gained by studying the exposure plots in Figure 7.11. Of primary interest are those frequency components closest to the zero frequency term for they contribute the most to the visibility of low frequency patterns.

Figure 7.11(a) is the same as that for the regular grid case on page 176 shrunk in the vertical direction, while (b) is shrunk by a factor of two in both directions. The horizontal banding in the spatial domain is evidenced by the strong vertically oriented frequency components indicated by arrows in (a). While the low frequency neighbors are symmetrically arranged in (b), all are lower in frequency. The success of the compensated array (c) is supported in the frequency domain by the radially symmetric arrangement of frequency samples.

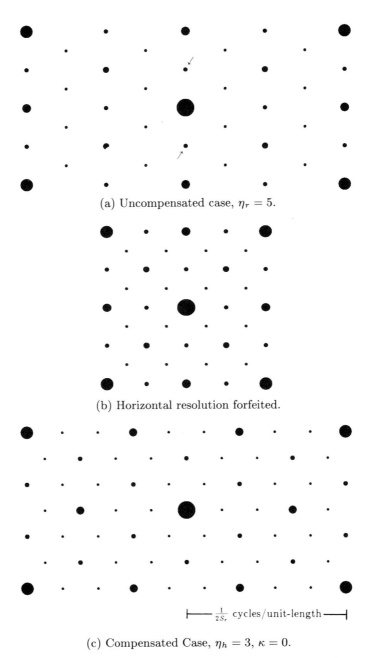

(a) Uncompensated case, $\eta_r = 5$.

(b) Horizontal resolution forfeited.

$\longmapsto \frac{1}{2S_r}$ cycles/unit-length $\longrightarrow$

(c) Compensated Case, $\eta_h = 3$, $\kappa = 0$.

Figure 7.11: Comparison of Exposure Plots for $\alpha_r = \frac{1}{2}$.

7.4 Examples for $\alpha_r = \frac{1}{6}$

The number of ternary subdivisions, κ, for the value of the rectangular aspect ratio, $\alpha_r = \frac{1}{6}$, is 1. To maintain a number of gray levels equal to that of the last section, a hexagonal threshold array of order $\eta_h = 2$ is used. Figure 7.12 shows the compensated threshold array to be used in this section. The total number of elements is $Z = 2(3)^{3+1} = 54$; the circled values are from the original hexagonal threshold array.

An aspect ratio of $\frac{1}{6}$ corresponds to a display which has the ability to resolve 6 times as many pixels in the S direction than in the L direction. This time 4 approaches are compared: the uncompensated rectangular ordered dither with $\eta = 5$, the case of sacrificing resolution for grid symmetry, the threshold array used in the last section for compensate for grids in the range $\frac{1}{3} < \alpha_r \leq 1$, with $\eta_h = 3$ and $\kappa = 0$, and finally the correctly compensated threshold array with $\eta_h = 2$ and $\kappa = 1$.

All four cases on the gray scale ramp are shown in Figure 7.13. The horizontal banding in (a) has become even more pronounced this extreme aspect ratio, as does the coarseness of (b). Horizontal banding even begins to appear in (c). In Figure 7.13(d), the asymmetries have been correctly compensated.

We've already seen the first two cases at the beginning of this chapter for the scanned picture in Figure 7.1(a) (page 188) and Figure 7.2 (page 188). The remaining two are illustrated in Figure 7.14. The uncorrected synthesize image seen in Figure 7.1(b) is also shown with proper threshold array in Figure 7.15.

As in the last section, the exposure plots provide much insight. All four cases conveniently fit on the one page in Figure 7.16. The plots in (a) and (b) are even more compressed than those for $\alpha_r = \frac{1}{2}$ in Figure 7.11, which follows the even greater low frequency artifacts in the spatial domain.

Recall that the threshold array with $\eta_h = 3$ and $\kappa = 0$ produced a composite Fourier transform with symmetrically arranged frequency components around the zero frequency center for the case of $\alpha_r = \frac{1}{2}$. In Figure 7.16(c) the location of the closest frequency components to the zero frequency center are also symmetrically arranged for $\alpha_r = \frac{1}{6}$, however, the horizontal banding is due to the greater concentration of energy in the vertical components indicated with arrows.

40

45 27 54 ③30 12 39

38 20 27 ⑤32 14 41 23 50 ⑧35 17 44

40 22 49 ⑦34 16 43 25 52 ①28 10 37 19 46 ④31 13 40

39 21 48 ⑥33 15 42 24 51 ⑨36 18 45

44 26 53 ②29 11 38

40

Figure 7.12: Compensated Threshold Array for $\frac{1}{9} < \alpha_r \le \frac{1}{3}$.
$\eta_h = 2$, $\kappa = 1$, $Z = 54$.
Circled values correspond to the original threshold array.

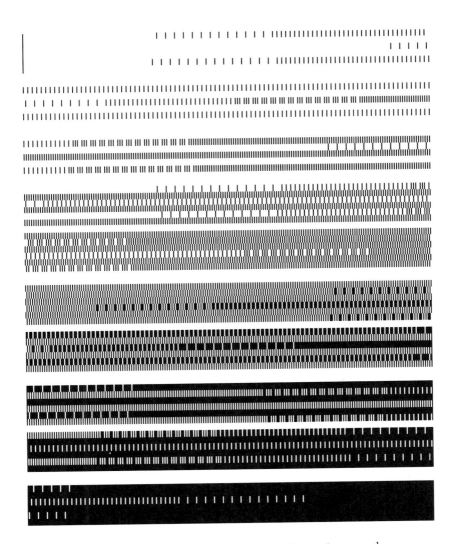

Figure 7.13: Comparison of Gray Scale Ramp for $\alpha_r = \frac{1}{6}$.
(a) Uncompensated Rectangular Ordered Dither, $\eta_r = 5$.

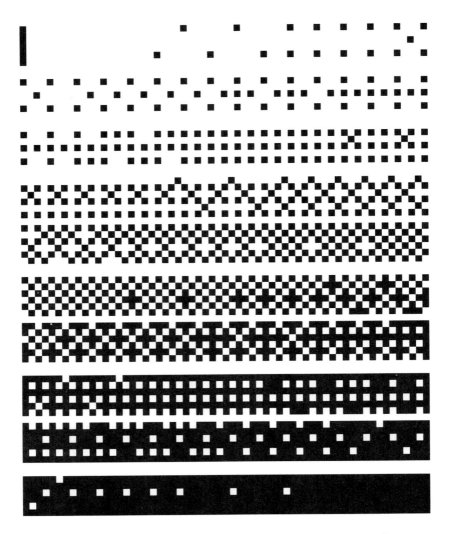

Figure 7.13:   Comparison of Gray Scale Ramp for $\alpha_r = \frac{1}{6}$.
(b) $\eta_r = 5$, sacrificing resolution for symmetry.

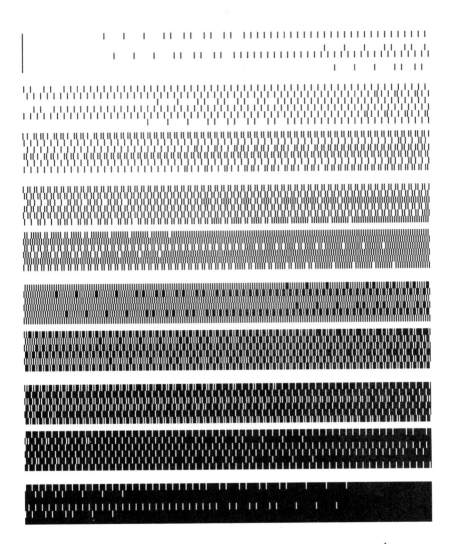

Figure 7.13: Comparison of Gray Scale Ramp for $\alpha_r = \frac{1}{6}$.
(c) Threshold Array, $\eta_h = 3$, $\kappa = 0$ compensated for the range $\frac{1}{3} < \alpha_r \le 1$.

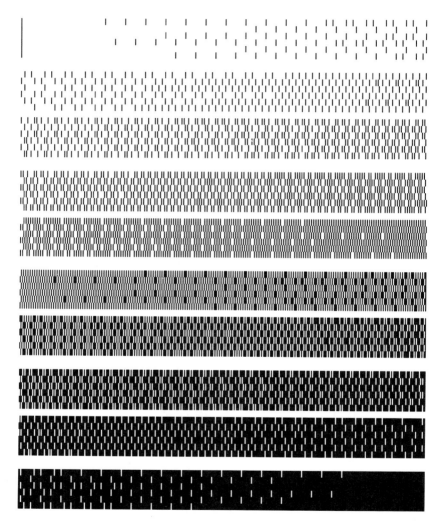

Figure 7.13: Comparison of Gray Scale Ramp for $\alpha_r = \frac{1}{6}$.
(d) Correctly Compensated Threshold Array, $\eta_h = 2$, $\kappa = 1$.

Figure 7.14: Comparison of Scanned Picture for $\alpha_r = \frac{1}{6}$.
(a) Threshold Array, $\eta_h = 3$, $\kappa = 0$ compensated for the range $\frac{1}{3} < \alpha_r \leq 1$.

Figure 7.14: Comparison of Scanned Picture for $\alpha_r = \frac{1}{6}$.
(b) Correctly Compensated Threshold Array, $\eta_h = 2$, $\kappa = 1$.

Figure 7.15: Synthesized Image with $\eta_h = 2$, $\kappa = 1$, and $\alpha_r = \frac{1}{6}$. (Compare with image on page 189.)

The exposure plot for the correctly compensated case in Figure 7.16(d) reflects the symmetry seen in the spatial domain. The frequency components are in the same locations as in case (c), however, the energy in those surrounding the zero frequency term is uniformly distributed.

7.5 Crossover Points

Equations (7.1) and (7.2) on page 196 for hexagonal grids and equations (7.3) and (7.4) on page 197 for rectangular grids precisely define the ranges over which particular values of κ will yield the most symmetric threshold arrays. This section addresses the nature of compensated ordered dither patterns on grids with aspect ratio values on the boundaries of those ranges.

7.5.1 Examples for $\alpha_r = 1$

The ranges in the above equations include regular hexagonal and rectangular grids. As was seen in section 7.1, hexagonal grids with aspect ratios $\alpha_h = \frac{2}{\sqrt{3}}$ or $\alpha_h = 2\sqrt{3}$ use the regular ordered dither threshold arrays unchanged. However, equation (7.3) suggests a threshold array formed from a hexagonal array with $\kappa = 0$ for use on regular rectangular grid ($\alpha_r = 1$).

Using the threshold array from Figure 7.8 (page 200), the gray scale ramp, scanned image, and composite Fourier transform are displayed in Figure 7.17. This output should be compared to that produced by ordered dither for regular rectangular grids in section 6.2. While some of the patterns produced in Figure 7.17(a) are radially symmetric, many suffer from some vertical banding. If the threshold array were generated from a hexagonal grid of the second kind (on the other side of the $\alpha_r = 1$ crossover point), the banding would be in the horizontal direction.

In section 6.2 it was pointed out that that since hexagonal grids with $\alpha_h = 2$ are actually regular rectangular grids, rectangular ordered dither arrays can be applied if rotated 45°. The hexagonally subsampled half of a regular rectangular grid will have an aspect ratio of 2.

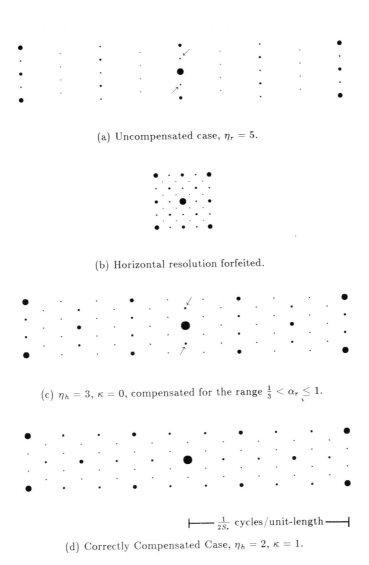

(a) Uncompensated case, $\eta_r = 5$.

(b) Horizontal resolution forfeited.

(c) $\eta_h = 3$, $\kappa = 0$, compensated for the range $\frac{1}{3} < \alpha_r \leq 1$.

$\longmapsto \frac{1}{2S_r}$ cycles/unit-length $\longmapsto$

(d) Correctly Compensated Case, $\eta_h = 2$, $\kappa = 1$.

Figure 7.16: Comparison of Exposure Plots for $\alpha_r = \frac{1}{6}$.

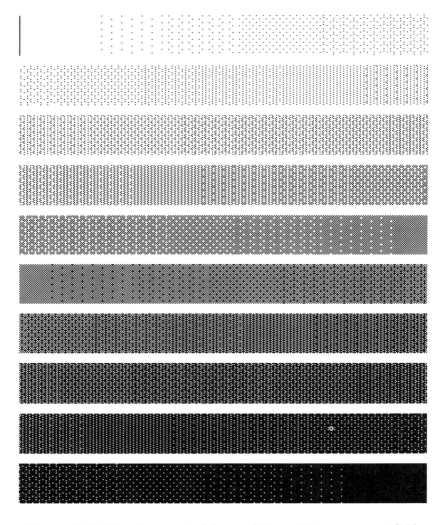

Figure 7.17: The Compensated Array of Figure 7.8 on an $\alpha_r = 1$ Grid.
(a) Gray Scale Ramp.

Figure 7.17: The Compensated Array of Figure 7.8 on an $\alpha_r = 1$ Grid.
(b) Scanned Image.

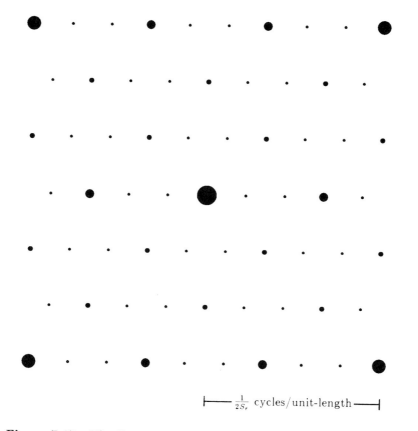

$\longmapsto \frac{1}{2S_r}$ cycles/unit-length $\longmapsto$

Figure 7.17: The Compensated Array of Figure 7.8 on an $\alpha_r = 1$ Grid.
(c) Composite Fourier Transform.

If a rectangular ordered dither array for *any* η_r is rotated 45° and applied to this hexagonal subset with the in between elements determined as in section 7.2, the result is *a rectangular ordered dither array of order $\eta_r + 1$!* So all of the threshold arrays in Chapter 6 are a special case of the generalized techniques of this chapter.

7.5.2 Examples for $\alpha_r = \frac{1}{3}$

The observation above is a surprising theoretical discovery for $\alpha_r = 1$. Rotating rectangular ordered dither arrays can be applied to other crossover points as well. Figure 2.5 on page 25 revealed that a hexagonal grid with aspect ratio $\alpha_h = 2$ is exactly as symmetric as a regular rectangular grid with $\alpha_r = 1$, at least in terms of covering efficiency, $E(\alpha)$. So it appears that either a hexagonal ordered dither array or a rectangular array rotated by 45° can be successfully applied to hexagonally subsampled grids with an effective aspect ratio of $\alpha_h = 2$.

In this section, the rectangular crossover point at $\alpha_r = \frac{1}{3}$ will be explored. Figure 7.18 illustrates how a threshold array at this aspect ratio can be generated. The underlined values are the hexagonal subsamples of the rectangular grid, and the circled values correspond to one ternary subsampling of the hexagonal subsamples. These circled subsamples have an aspect ratio of $\alpha_h = 2$; the values assigned are those of a rectangular ordered dither threshold array with $\eta_r = 3$ rotated 45°. The in between values are then assigned in the usual way. The value of $\eta_r = 3$ was chosen so as to produce a threshold array size $(Z = 48)$ as close as possible to that of the other two arrays to be compared.

The effect of halftoning with

(a) the threshold array of Figure 7.8 (page 200) with $\eta_h = 3$, $\kappa = 0$, and $Z = 54$ compensated for the range $\frac{1}{3} < \alpha_r \leq 1$,

(b) the threshold array of Figure 7.12 (page 210) with $\eta_h = 2$, $\kappa = 1$, and $Z = 54$ compensated for the range $\frac{1}{9} < \alpha_r \leq \frac{1}{3}$, and

(c) the threshold array of Figure 7.18 with $\eta_r = 3$, $\kappa = 1$, and $Z = 48$

is compared in Figure 7.19 for the gray scale ramp, and Figure 7.20 for the scanned picture.

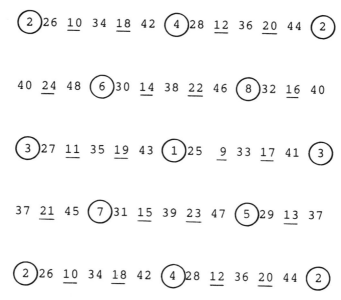

Figure 7.18: Threshold Array designed for $\alpha_r = \frac{1}{3}$.
Circled values correspond to the rectangular array with $\eta_r = 3$, rotated 45°.
Underlined values are the result of one ternary replication.

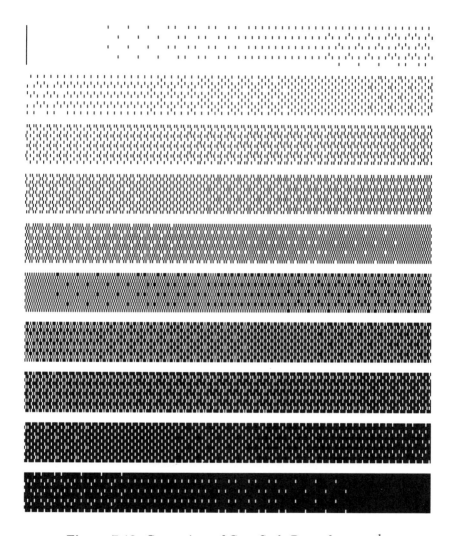

Figure 7.19: Comparison of Gray Scale Ramp for $\alpha_r = \frac{1}{3}$.
(a) Threshold Array, $\eta_h = 3$, $\kappa = 0$ compensated for the range $\frac{1}{3} < \alpha_r \leq 1$.

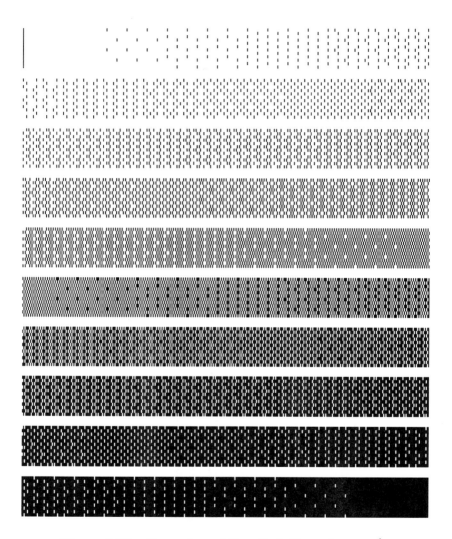

Figure 7.19:   Comparison of Gray Scale Ramp for $\alpha_r = \frac{1}{3}$.
(b) Threshold Array, $\eta_h = 2$, $\kappa = 1$ compensated for the range $\frac{1}{9} < \alpha_r \leq \frac{1}{3}$.

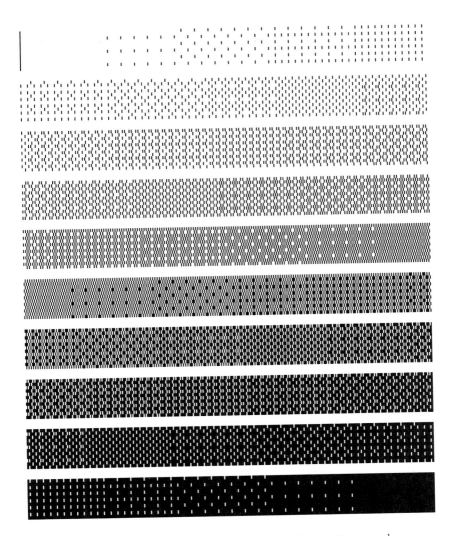

Figure 7.19: Comparison of Gray Scale Ramp for $\alpha_r = \frac{1}{3}$.
(c) Threshold Array of Figure 7.18.

Figure 7.20: Comparison of Scanned Picture for $\alpha_r = \frac{1}{3}$.
(a) Threshold Array, $\eta_h = 3$, $\kappa = 0$ compensated for the range $\frac{1}{3} < \alpha_r \leq 1$.

Figure 7.20: Comparison of Scanned Picture for $\alpha_r = \frac{1}{3}$.
(b) Threshold Array, $\eta_h = 2$, $\kappa = 1$ compensated for the range $\frac{1}{9} < \alpha_r \leq \frac{1}{3}$.

Figure 7.20: Comparison of Scanned Picture for $\alpha_r = \frac{1}{3}$.
(c) Threshold Array of Figure 7.18.

Once again, much can be learned by looking at the composite Fourier transforms for this case in Figure 7.21. The slight horizontal and vertical artifacts observed in the spatial domain for cases (a) and (b) are of equal but opposite magnitude at the crossover point. Their associated Exposure Plots reveal this asymmetry with the guilty frequency components highlighted with arrows. The Exposure Plot in Figure 7.21(c) does not show any sign of such asymmetry, however, it possesses horizontal and vertical frequency components that are lower in frequency than those found in (a) or (b). It appears that the optimum choice is a toss-up at such crossover points.

7.6 Summary

This concludes the investigation of the point process of ordered dither.

Chapter 5 reviewed the relatively simple problem of generating clustered-dot halftone patterns on both rectangular and hexagonal grids. The more involved problem of dispersed-dot ordered dither has been generalized for *all* semiregular grids. Chapter 6 developed the basic theory with the Method of Recursive Tessellation to generate threshold arrays for regular grids, both rectangular and hexagonal, with odd or even periods. Here in Chapter 7, a general rule for creating correctly compensated threshold arrays for any asymmetric aspect ratio has been established.

The use of exposure plots to represent composite Fourier transforms of all possible patterns produced by a given threshold array on a particular grid adds depth to our understanding of the ordered dither process. It reveals the low frequency clustering of energy for clustered dot patterns, the high frequency distribution for dispersed-dot patterns, and culprits of asymmetry in cases where such artifacts exist.

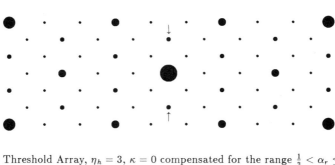

(a) Threshold Array, $\eta_h = 3$, $\kappa = 0$ compensated for the range $\frac{1}{3} < \alpha_r \leq 1$.

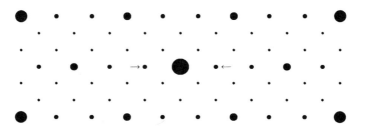

(b) Threshold Array, $\eta_h = 2$, $\kappa = 1$ compensated for the range $\frac{1}{9} < \alpha_r \leq \frac{1}{3}$.

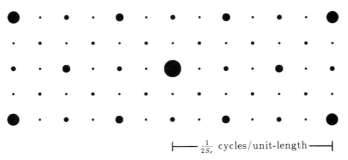

$\vdash\!\!\!\!-\!\!\!\!-\frac{1}{2S_r}$ cycles/unit-length $-\!\!\!\!-\!\!\!\!\dashv$

(c) Threshold Array of Figure 7.18.

Figure 7.21: Comparison of Exposure Plots for $\alpha_r = \frac{1}{3}$.

Chapter 8

Dithering with Blue Noise

The nature of various types of noise is often described by a color name. The most well known example is "white noise", so named because its power spectrum is flat across all frequencies, much like the visible frequencies in white light. "Pink noise" is used to describe low frequency white noise, the power spectrum of which is flat out to some finite high frequency limit. There is even the curious case of "brown noise", named for the spectrum associated with Brownian motion [25]. Introduced in this chapter is "blue noise", the high frequency complement of pink noise, which will be shown to be important in the generation of good digital halftones.

Unlike any of the halftoning techniques presented thus far in this book, schemes which generate blue noise are neighborhood operations. As in Chapter 4, which explored dithering with white noise, blue noise patterns are aperiodic and radially symmetric. Although white noise patterns do not suffer from the correlated periodicity of ordered dither, the fact that they possess energy at very low frequencies results in a grainy appearance. Blue noise patterns enjoy the benefits of aperiodic, uncorrelated structure without low frequency graininess.

The important algorithm known as error diffusion is closely examined and with some variation is found to be a good blue noise generator. This algorithm is examined for both rectangular and hexagonal grids.

After the success of hexagonal grids for the case of ordered dither, it is surprising to learn, as will be theoretically established, that a rectangular grid is the superior choice for dithering with blue noise.

8.1   Principal Wavelength

Consider the problem of rendering a fixed gray level, g, with binary pixels on regular rectangular or hexagonal grids. A goal is to distribute the binary pixels as homogeneously as possible. These pixels would be separated by an average distance in two dimensions. This distance is called the *Principal Wavelength*, and would have the value

$$\lambda_g = \begin{cases} |\mathbf{v}|/\sqrt{g} & g \le \frac{1}{2} \\ |\mathbf{v}|/\sqrt{1-g} & g > \frac{1}{2} \end{cases} \tag{8.1}$$

where $|\mathbf{v}| = |\mathbf{v_1}| = |\mathbf{v_2}| = S$, and $\mathbf{v_1}$ and $\mathbf{v_2}$ are the spatial sampling vectors defined in section 2.1.1. $|\mathbf{v_1}| = |\mathbf{v_2}|$ because we are considering regular grids only.

Since the distribution is assumed to be homogeneous, the corresponding power spectrum would be radially symmetric. The principal wavelength would be manifested as the *Principal Frequency*,

$$f_g = \begin{cases} \sqrt{g}\,|\mathbf{u}| & g \le \frac{1}{2} \\ \sqrt{1-g}\,|\mathbf{u}| & g > \frac{1}{2} \end{cases} \tag{8.2}$$

where $|\mathbf{u}| = |\mathbf{u_1}| = |\mathbf{u_2}|$, and $\mathbf{u_1}$ and $\mathbf{u_2}$ are the frequency baseband replication vectors. Recalling that for a regular rectangular grid, $L = S$, and for a regular hexagonal grid of the first kind, $L = \frac{\sqrt{3}}{2}S$. Upon combining these with the relations of Equations (3.19) and (3.21) (page 48),

$$|\mathbf{u}| = \begin{cases} \dfrac{1}{S} & \text{for rectangular grids.} \\ \dfrac{2}{\sqrt{3}S} & \text{for hexagonal grids.} \end{cases} \tag{8.3}$$

We can now plot f_g as a function of g in Figure 8.1 for (a) rectangular and (b) hexagonal grids. These plots reveal an amazing shortcomings

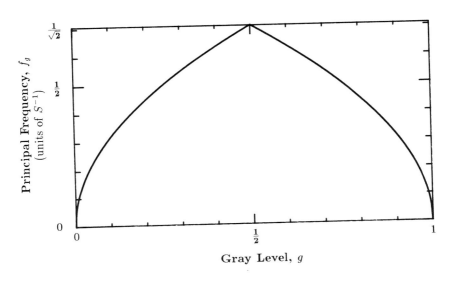

(a) Regular Rectangular Grid.

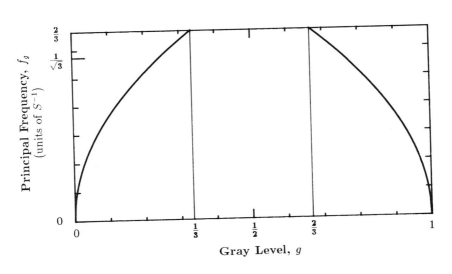

(b) Regular Hexagonal Grid.

Figure 8.1: Principal Frequency, f_g, as a function of g.

of hexagonal grids; they cannot support a principal frequency for $\frac{1}{3} <$ $g < \frac{2}{3}$! Figure 8.2(b) illustrates the highest frequency which can exist on a hexagonal grid, $f_r/S^{-1} = \frac{2}{3}$, for either a pattern for $g = \frac{1}{3}$ or its complement, $g = \frac{2}{3}$. On a rectangular grid (Figure 8.2(a)), the checkerboard pattern for gray level, $g = \frac{1}{2}$, and frequency, $f_r/S^{-1} = \frac{1}{\sqrt{2}}$, can most definitely be supported. These spatial patterns correspond to the high frequency corners of the frequency baseband.

For gray levels in the range $\frac{1}{3} < g < \frac{2}{3}$ on a hexagonal grid, pixels must be grouped together resulting in frequencies lower than $f_r/S^{-1} = \frac{2}{3}$. A principal frequency exists for all gray levels on rectangular grids.

A well formed binary dither pattern rendering of a fixed gray level should consist of an isotropic field of binary pixels with an average separation of λ_g. This average separation should vary in an uncorrelated manner, but the wavelengths of this variation must not be significantly longer that λ_g. The failure of dithering with white noise was due to the presence of long wavelengths.

Figure 8.3 depicts these well formed dither pattern characteristics in the frequency domain. The radially averaged power spectrum (defined in section 3.2.2) of a fixed gray level, g, has 3 important features. First, its peak should be at the principal frequency for that gray level, f_g. This frequency marks a sharp transition region below which little of no energy exists. And finally, the uncorrelated high frequency fluctuations are characterized by high frequency white noise, or "blue noise".

The visually pleasing nature of some of the patterns generated by the error diffusion algorithm to be presented in the next section can be attributed to a spectral signature as just described. However, several shortcomings exist in this algorithm. In section 8.3, error diffusion enhanced with certain stochastic perturbations is found to be a good blue noise generator.

The presence of significant low frequency energy is responsible for the visibility of disturbing artifacts in halftoning patterns. For dispersed-dot ordered dither, half of the total possible gray level patterns available for a given threshold array have low frequency components which correspond to wavelengths of the size of the threshold period. For good blue noise processing, the lowest frequency is essentially f_g. The negative feedback of error diffusion acts as a low frequency inhibitor.

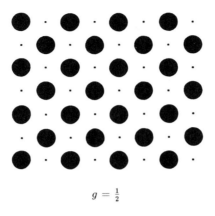

$$g = \tfrac{1}{2}$$

(a) Rectangular Grid, $f_g/S^{-1} = \tfrac{1}{\sqrt{2}}$.

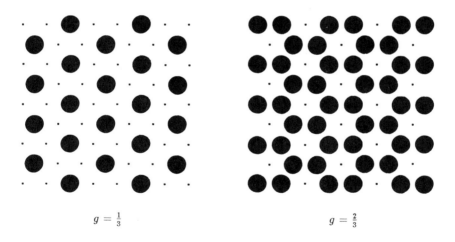

$$g = \tfrac{1}{3} \qquad\qquad g = \tfrac{2}{3}$$

(b) Hexagonal Grid, $f_g/S^{-1} = \tfrac{2}{3}$.

Figure 8.2: Patterns with the highest possible Spatial Frequency (Corresponding to the corners of the spectral baseband).

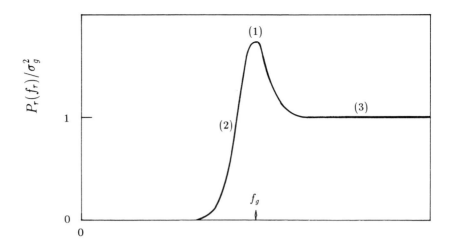

Radial Frequency, f_r (units of S^{-1})

Figure 8.3: Spectral Characteristics of a Well Formed Dither Pattern.

1. Low frequency cutoff at principal frequency.
2. Sharp transition region.
3. Flat high frequency "Blue Noise" region.

Figure 8.4: Point Processes.

8.2 The Error Diffusion Algorithm

Up to this point, all of the halftone processes considered could be modeled as shown in Figure 8.4. All were point processes, that is, the output depends only on the current input pixel. Pixels of the continuous-tone or m-ary digital image, $J[\mathbf{n}]$, are simply compared with a threshold to determine the state of the output pixels, $I[\mathbf{n}]$.

The error diffusion algorithm, first introduced by Floyd and Steinberg in 1975 [20,21], requires neighborhood operations and is thus more computationally intensive. It is currently the most popular neighborhood halftoning process and has received considerable attention in spite of some shortcomings. A generic form of this algorithm is graphically illustrated in Figure 8.5.

The threshold in this case is fixed at $\frac{1}{2}$ where the input, $J[\mathbf{n}]$, varies as usual from $g = 0$ (white) to $g = 1$ (black). The resulting binary output value of 0 or 1 is compared with the original gray level value. The difference is suitably called the "error" for location $\mathbf{n}$. The signal consisting of past error values is passed through an error filter, $e[\mathbf{n}]$, to produce a correction factor to be added to future input values. Errors are thus "diffused" over a weighted neighborhood determined by $e[\mathbf{n}]$.

Figure 8.6 summarizes error filter impulse responses promoted in the literature. Note that in all cases the values are deterministic and sum to 1 so that errors are neither amplified nor reduced. The first three listed are designed for rectangular grids and will be examined in both the spatial and frequency domain in this section. The error filter in (d) is intended for use on a hexagonal grid; it will be similarly addressed in section 8.4.

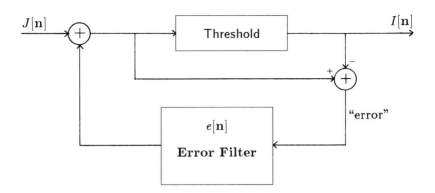

Figure 8.5: The Error Diffusion Algorithm.

$$\left(\frac{1}{16}\times\right) \qquad \begin{array}{ccc} & \bullet & 7 \\ 3 & 5 & 1 \end{array}$$

(a) Floyd and Steinberg (1975) [20,21].

$$\left(\frac{1}{48}\times\right) \qquad \begin{array}{ccccc} & & \bullet & 7 & 5 \\ 3 & 5 & 7 & 5 & 3 \\ 1 & 3 & 5 & 3 & 1 \end{array}$$

(b) Jarvis, Judice and Ninke (1976) [36].

$$\left(\frac{1}{42}\times\right) \qquad \begin{array}{ccccc} & & \bullet & 8 & 4 \\ 2 & 4 & 8 & 4 & 2 \\ 1 & 2 & 4 & 2 & 1 \end{array}$$

(c) Stucki (1981) [80].

$$\left(\frac{1}{200}\times\right) \qquad \begin{array}{ccccccc} & & & \bullet & & 32 & \\ 12 & & 26 & & 30 & & 16 \\ & 12 & & 26 & & 12 & \\ 5 & & 12 & & 12 & & 5 \end{array}$$

(d) Stevenson and Arce (1985) [76].

Figure 8.6: Error Filters reported in the Literature.
(a), (b) and (c) are for rectangular grids.
(d) is for hexagonal grids.

("•" represents the origin.)

The original error filter or set of "weights" suggested by Floyd and Steinberg is shown in Figure 8.6(a). They argued that a filter with four elements was the smallest number that could produce "good" results. The values were chosen to particularly assure the checkerboard pattern at middle gray.

Figure 8.7 shows the effect of error diffusion with the Floyd and Steinberg filter for (a) the gray scale ramp, (b) scanned picture, and (c) the synthesized image. The reason for the popularity of this algorithm is clear; several gray levels are represented by pleasingly isotropic, structureless distributions of dots.   However, some shortcomings are also apparent:

1. Correlated artifacts in many of the gray level patterns. This can best be seen in the gray scale ramp.

2. Directional hysteresis due to the raster order of processing. This artifact is most apparent in very light and very dark patterns. Note the light regions of the sky in the scanned picture, the highlight in the egg shaped object and the dark squares of the synthesized image.

3. Transient behavior near edges or boundaries. A good example of this is in the rendering of the fixed background at the top and left edges of the synthesized image.

The radially averaged power spectrum, $P_r(f_r)$, and anisotropy measure, $s^2(f_r)/P_r^2(f_r)$ (described in section 3.2.2) are plotted for selected gray levels in Figure 8.8 on pages 246 through 252. In each case here and throughout this chapter, the principal frequency, f_g, is marked with a small diamond on the frequency axis of the power spectrum.

The lack of symmetry in these patterns is strongly acknowledged in the anisotropy measure, especially for (d) $g = \frac{1}{4}$ and (e) $g = \frac{1}{2}$. Recall that the "background noise limit" due to the spectral estimation method is -10 dB (equation (3.33)) indicated by a reference line at that level. Any measure greater than 0 dB at any frequency indicates an especially anisotropic pattern; at such a level the sample variance is greater than the square of the average.

Figure 8.7: Error Diffusion with the Floyd and Steinberg Filter. (a) Gray Scale Ramp, $\alpha_r = 1$.

Figure 8.7:   Error Diffusion with the Floyd and Steinberg Filter.
(b) Scanned Picture, $\alpha_r = 1$.

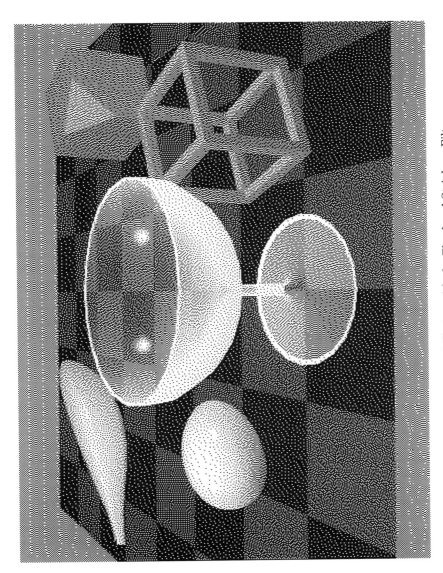

Figure 8.7: Error Diffusion with the Floyd and Steinberg Filter. (c) Synthesized Image, $\alpha_r = 1$.

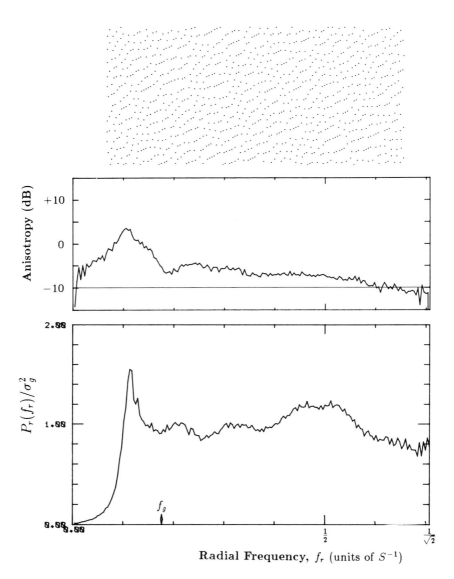

Figure 8.8: Radial Spectra for the Floyd and Steinberg Filter.
(a) $\boxed{g = \frac{1}{32}}$, $f_g/S^{-1} \approx .1768$, $\sigma_g^2 \approx .0303$

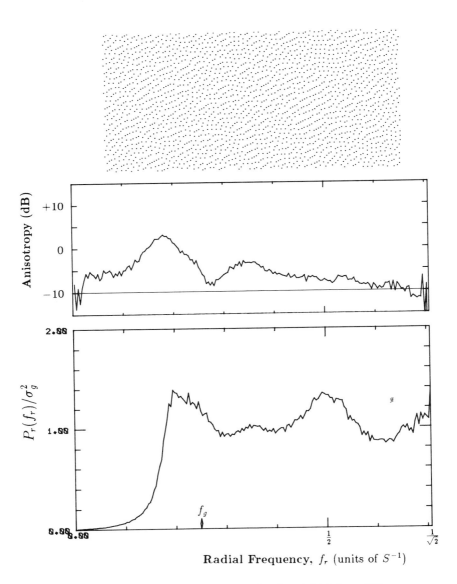

Figure 8.8: Radial Spectra for the Floyd and Steinberg Filter.
(b) $\boxed{g = \frac{1}{16}}$, $f_g/S^{-1} = .25$, $\sigma_g^2 \approx .0586$

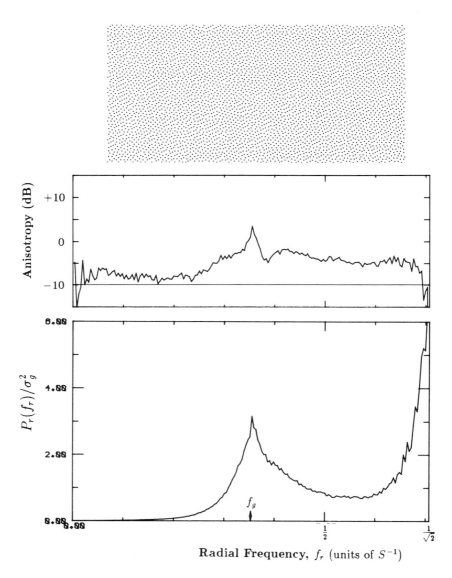

Figure 8.8: Radial Spectra for the Floyd and Steinberg Filter. (c) $\boxed{g = \frac{1}{8}}$, $f_g/S^{-1} \approx .3495$, $\sigma_g^2 \approx .1094$

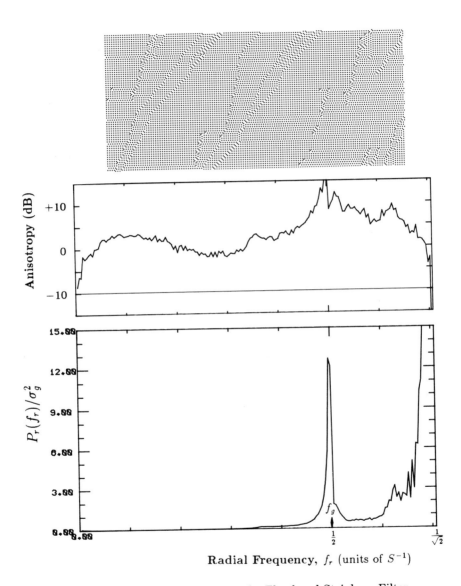

Figure 8.8: Radial Spectra for the Floyd and Steinberg Filter.
(d) $\boxed{g = \frac{1}{4}}$, $f_g/S^{-1} = .5$, $\sigma_g^2 = .1875$

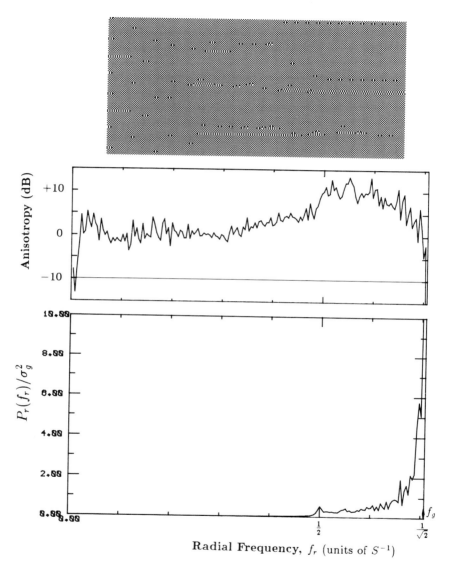

Figure 8.8: Radial Spectra for the Floyd and Steinberg Filter.
(e) $\boxed{g = \frac{1}{2}}$, $f_g/S^{-1} \approx .7071$, $\sigma_g^2 = .25$

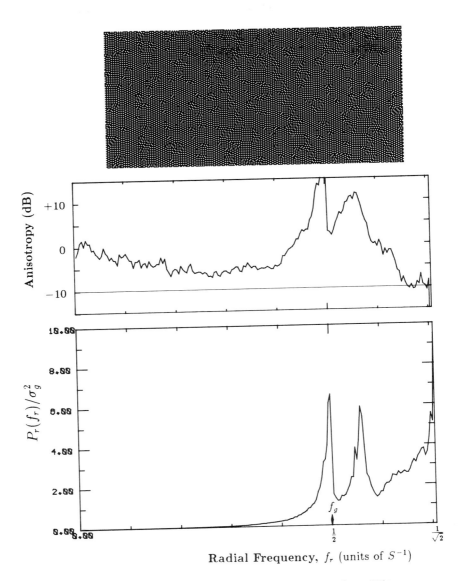

Figure 8.8: Radial Spectra for the Floyd and Steinberg Filter.
(f) $\boxed{g = \frac{3}{4}}$, $f_g/S^{-1} = .5$, $\sigma_g^2 = .1875$

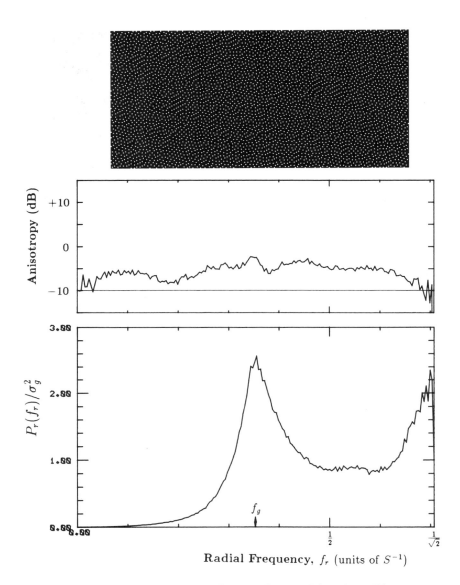

Figure 8.8:   Radial Spectra for the Floyd and Steinberg Filter.
(g) $\boxed{g = \frac{7}{8}}$, $f_g/S^{-1} \approx .3536$, $\sigma_g^2 \approx .1094$

8.2.1 Filters with 12 Weights

In 1976 Jarvis, Judice and Ninke [36] reported an error filter with the 12 elements shown in Figure 8.6(b). Although their technique was identical to error diffusion, they called it the "Minimum Average Error" algorithm. Output from error diffusion with this filter is shown in Figure 8.9. The larger filter size does reduce some of the artifacts seen with the 4 element filter of Floyd and Steinberg, but directional hysteresis in the very dark and light regions has increased, and pixels are clustered together more in the middle gray regions.

It also sharpens the picture more. The amount of sharpening may or may not be to the degree desired. In fact, Stucki [80] argues that it is objectionable and uses an additional filter to *inhibit* sharpening. *Some* symmetric sharpening is usually desirable, and is easily achieved by including a separate presharpening step (see Figure 1.3). This will be demonstrated in section 9.1.1 where the degree of sharpening can be precisely controlled.

The 12 element error filter used by Stucki is shown in Figure 8.6(c). For computational efficiency, the selected values are all powers of 2. The effect of error diffusion with this error filter is displayed in Figure 8.10, and is found to be quite similar to the images of Figure 8.9. In each case the directional nature of the inherent sharpening is particularly evident in the synthesized image under the glass-shaped object. In that image, the background which has a fixed value just under $g = \frac{1}{2}$, reveals high sensitivity to the slight difference between the two 12 weight filters.

The patterns generated for fixed gray levels with the Jarvis, *et al.* filter is examined in the frequency domain in Figure 8.11 (pages 260 through 264). While all gray levels shown still suffer some anisotropy, $g = \frac{1}{32}$, $\frac{1}{16}$, and $\frac{1}{8}$ show a stronger concentration of energy at the principal frequency, f_g, than those of Figure 8.8. The clustering of pixels at $g = \frac{1}{4}$ and $\frac{1}{2}$ results in significant energy at frequencies lower than f_g.

Figure 8.9: Error Diffusion with the Jarvis, *et al.* Filter.
(a) Gray Scale Ramp, $\alpha_r = 1$.

Figure 8.9: Error Diffusion with the Jarvis, *et al.* Filter.
(b) Scanned Picture, $\alpha_r = 1$.

Figure 8.9: Error Diffusion with the Jarvis, *et al.* Filter. (c) Synthesized Image, $\alpha_r = 1$.

Figure 8.10: Error Diffusion with the Stucki Filter.
(a) Gray Scale Ramp, $\alpha_r = 1$.

Figure 8.10:   Error Diffusion with the Stucki Filter.
(b) Scanned Picture, $\alpha_r = 1$.

Figure 8.10: Error Diffusion with the Stucki Filter. (c) Synthesized Image, $\alpha_r = 1$.

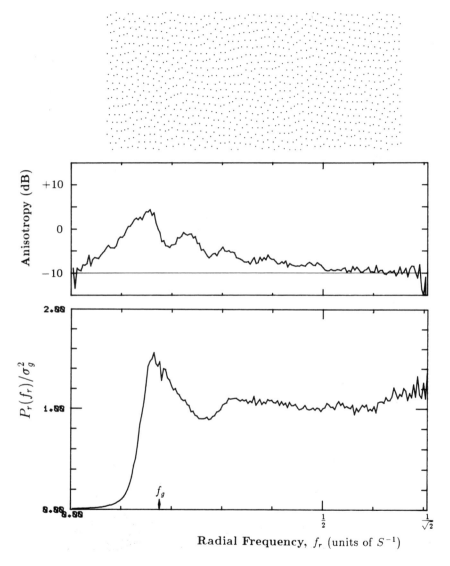

Figure 8.11: Radial Spectra for the Jarvis, *et al.* Filter.
(a) $\boxed{g = \frac{1}{32}}$, $f_g/S^{-1} \approx .1768$, $\sigma_g^2 \approx .0303$

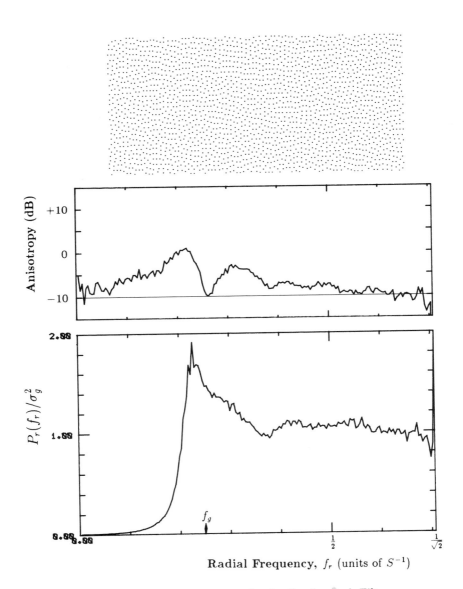

Figure 8.11: Radial Spectra for the Jarvis, *et al.* Filter.
(b) $\boxed{g = \frac{1}{16}}$, $f_g/S^{-1} = .25$, $\sigma_g^2 \approx .0586$

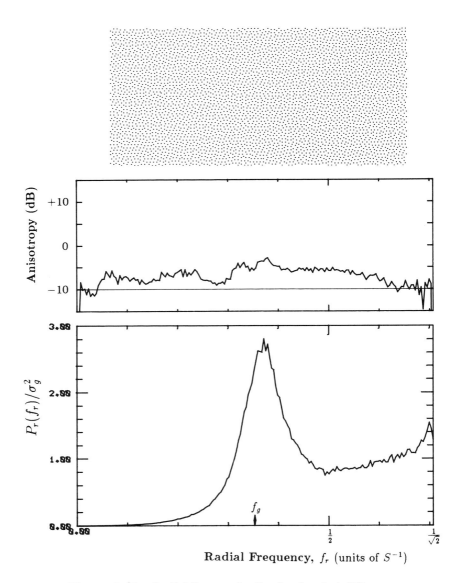

Figure 8.11: Radial Spectra for the Jarvis, *et al.* Filter.

(c) $\boxed{g = \frac{1}{8}}$, $f_g/S^{-1} \approx .3495$, $\sigma_g^2 \approx .1094$

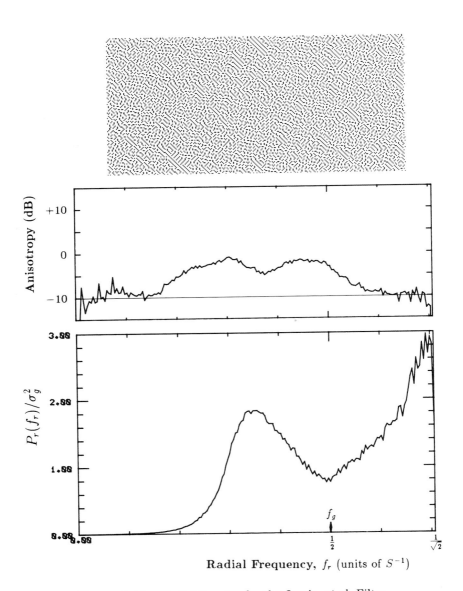

Figure 8.11: Radial Spectra for the Jarvis, *et al.* Filter.
(d) $\boxed{g = \frac{1}{4}}$, $f_g/S^{-1} = .5$, $\sigma_g^2 = .1875$

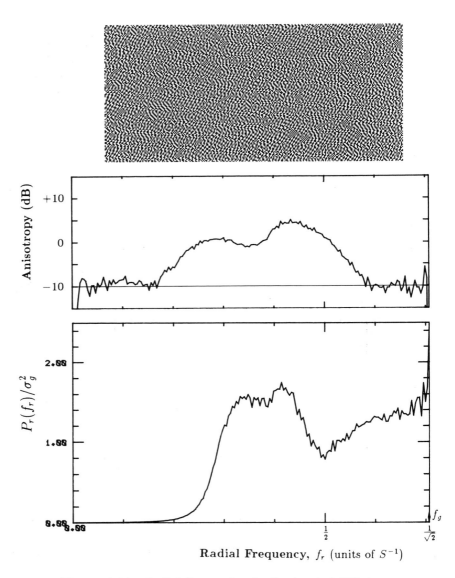

Radial Frequency, f_r (units of S^{-1})

Figure 8.11: Radial Spectra for the Jarvis, *et al.* Filter.
(e) $\boxed{g = \frac{1}{2}}$, $f_g/S^{-1} \approx .7071$, $\sigma_g^2 = .25$

8.3 Error Diffusion with Perturbation

The idea of "dithering" or perturbing a method in image processing to defeat visual artifacts of a regular and deterministic nature has been used before. Randomizing sampling grids [17] is one method used to reduce the aliasing effects of undersampled images. Allebach improved the classical screen by randomizing the centers of dot clusters [2,4] which eliminated the occurrence of morié patterns. And in the color printing industry, arbitrary or "irrational" clustered-dot screen angles have been digitally produced by employing random perturbations [23,64].

Proposed here are several modifications to the basic error diffusion algorithm graphically depicted in Figure 8.5. They are categorized in the following four areas.

Choice of Error Filter

An enormous number of choices are available for this, the deterministic part of the algorithm. An error filter can consist of weights of any

1. number,

2. position, and

3. value.

For computational efficiency, as small a filter as possible is preferred.

Threshold Perturbation

In 1983, Billotet-Hoffman and Bryngdahl [11] suggested using an ordered dither threshold array in place of the fixed threshold used in error diffusion. However, the resulting halftoned output differs little from conventional ordered dither. A modification to this idea would be to perturb a fixed threshold within a given maximum percentage with ordered dither and/or white noise.

So, additional parameters include:

1. Choice of period size of ordered dither.

2. Magnitude of ordered dither perturbation.

3. Magnitude of white noise perturbation.

Raster Direction

The directional artifacts seen in the examples of error diffusion are due largely to the traditional raster order of processing. Many choices of space filling curves to define the order of processing are possible. Although they did not call it error diffusion, Witten and Neal [88] demonstrated fairly good results by essentially using an error filter with *one* deterministic weight and processing all of the two-dimensional image data on a Peano curve (a type of fractal). While this particular approach imposes heavy demands on memory, the idea of using non-standard raster ordering should be tried.

One idea that breaks up the directionality of a normal raster without the expense of a full two-dimensional buffer is to process along a serpentine raster (see Figure 8.12). Neighborhood operations, in image processing hardware or software, buffer image data in full lines. So the choice of serpentine raster processing does not require any memory increase over a normal raster.

Stochastic Filter Perturbations

Along with threshold perturbations, random noise can be added to the elements or weights of the error filter. This idea was proposed by Schreiber [66] and demonstrated by Woo [89], but only on the 12 element filter of Jarvis, *et al.*.

The magnitude or range of additive noise can be adjusted for each element. The sum of all of the weights in the resulting stochastic filter should still be unity at all times. This condition can be met by pairing weights of comparable value. For each pair of weights a scaled random value is added to one and subtracted from the other.

A random value, χ, is generated by the method described in Chapter 4, with the adjusted uniform probability density function,

$$
p_\chi(\chi) = \begin{cases} \dfrac{a_i}{2} & \text{for} \quad -a_i < \chi < a_i \\ 0 & \text{otherwise} \end{cases} \tag{8.4}
$$

for each pair of weights, i, in the filter. Then at every image location for each pair of weights, $a_i\chi$ is added to the first weight, and subtracted from the second. The a_i's, expressed as a percentage of the smaller weight in the pair, are yet other adjustable parameters.

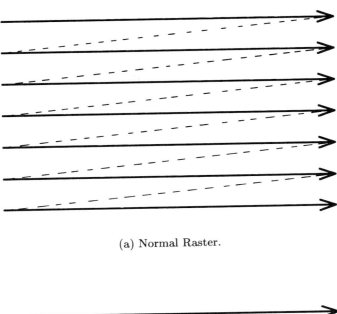

(a) Normal Raster.

(b) Serpentine Raster.

Figure 8.12: Two Processing Path Options.

The error filter can also be perturbed by randomizing the *positions* of the weights.

The number of adjustable parameters available to modify the basic error diffusion algorithm is tremendously large. Much is to be learned about the effect of each parameter used independently and in combination with others. Over a hundred combinations of these parameters were experimented with in this investigation; the examples that follow are a carefully selected representative sample.

8.3.1   One Weight

The most computationally inexpensive form of the error diffusion algorithm is one implemented with an error filter with one weight. In this case all of the error is diffused to only one location; no multiplication is necessary.

If the one weight is fixed to some predetermined location, the resulting patterns fail in a big way. This is demonstrated in Figure 8.13 where the position of the weight was fixed diagonally adjacent to the origin. Failure is at least as great for any other location choice, or when a serpentine raster is used.

However, when the *position* of the one weight is randomly determined over some finite set of candidate positions, a much more acceptable result emerges. The images of Figure 8.14 where produced with the position of the one weight selected with equal probability between only two candidate locations, immediately below and preceding the filter origin.

This example represents a broad class of parameter combinations. Effectively the same output results for any local neighborhood of candidate locations, as well as for a broad range of probability mass functions governing the selection of a location. Several combinations were tried without change in output, including the 4 Floyd and Steinberg locations using the values of their weights as the probability mass function for position selection. Even when 2 or 3 weights were randomly selected, no significant difference was seen.

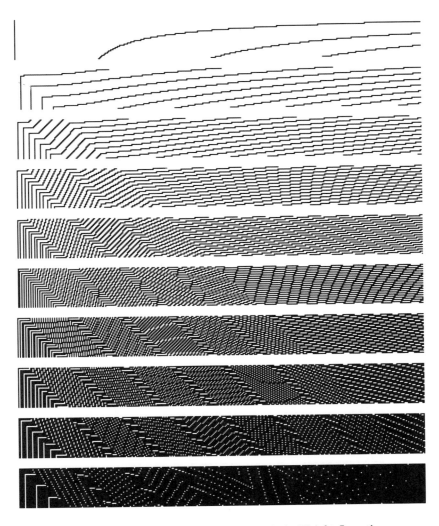

Figure 8.13: Failure of One Deterministic Weight Location.

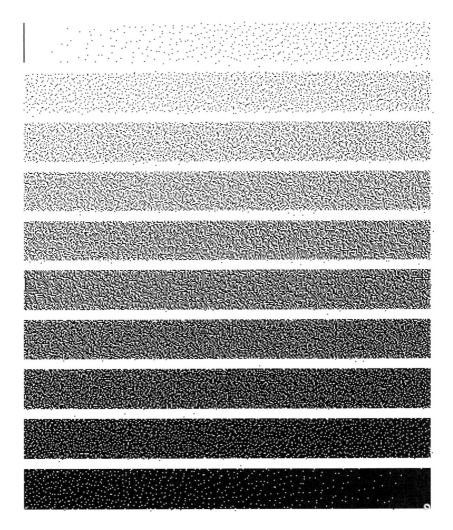

Figure 8.14: Effect of One Randomly Positioned Weight.
(a) Gray Scale Ramp, $\alpha_r = 1$.

Figure 8.14: Effect of One Randomly Positioned Weight.
(b) Scanned Picture, $\alpha_r = 1$.

The radially averaged power spectrum is displayed in Figure 8.15. The spectra for the various gray levels all reveal the desirable properties of

1. very low anisotropy,

2. flat blue noise region, and

3. cutoff at f_g.

The only feature that the spectra fall short of is that of a sharp transition region. This is most pronounced at $g = \frac{1}{2}$. The low frequency leakage is responsible for the grainy texture; the associated patterns can be called "light-blue noise". The suppression of frequencies below f_g is sufficient, however, to produce a vast improvement over the white noise images of Chapter 4.

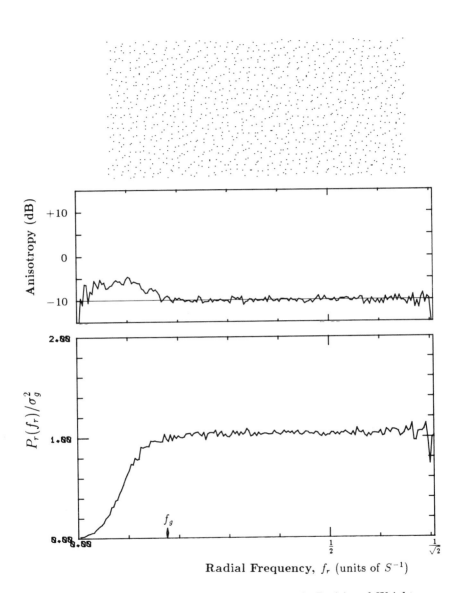

Figure 8.15: Radial Spectra for One Randomly Positioned Weight.
(a) $\boxed{g = \frac{1}{32}}$, $f_g/S^{-1} \approx .1768$, $\sigma_g^2 \approx .0303$

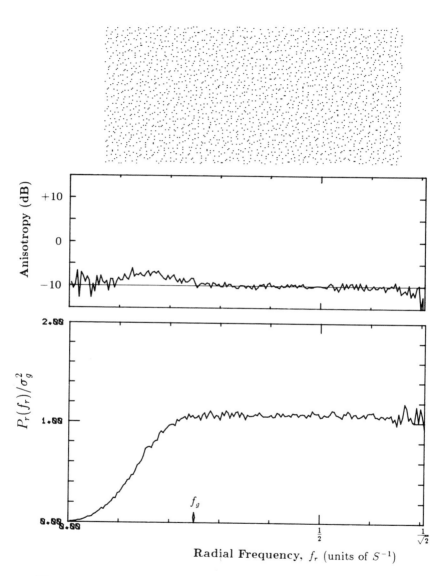

Figure 8.15:   Radial Spectra for One Randomly Positioned Weight.
(b) $\boxed{g = \frac{1}{16}}$, $f_g/S^{-1} = .25$, $\sigma_g^2 \approx .0586$

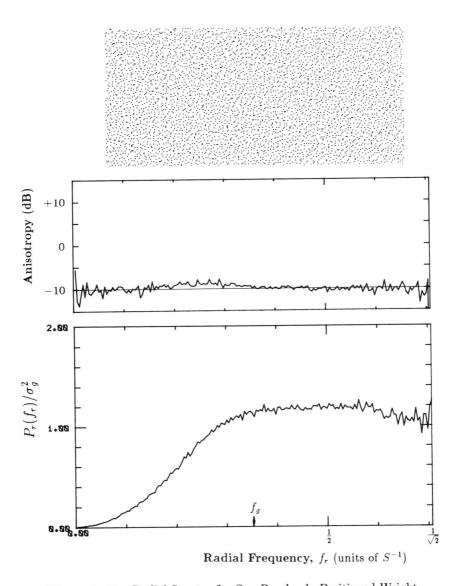

Figure 8.15: Radial Spectra for One Randomly Positioned Weight. (c) $\boxed{g = \frac{1}{8}}$, $f_g/S^{-1} \approx .3495$, $\sigma_g^2 \approx .1094$

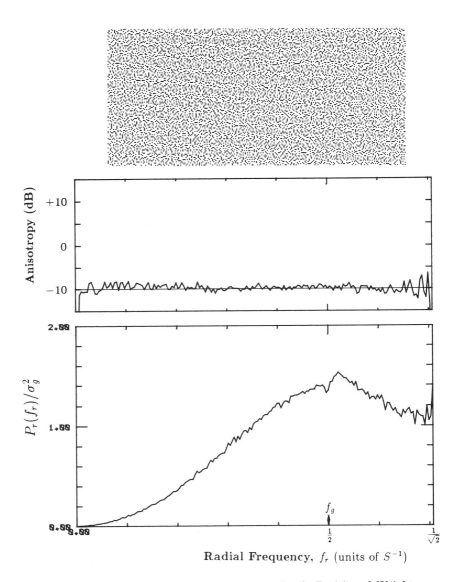

Figure 8.15: Radial Spectra for One Randomly Positioned Weight. (d) $\boxed{g = \frac{1}{4}}$, $f_g/S^{-1} = .5$, $\sigma_g^2 = .1875$

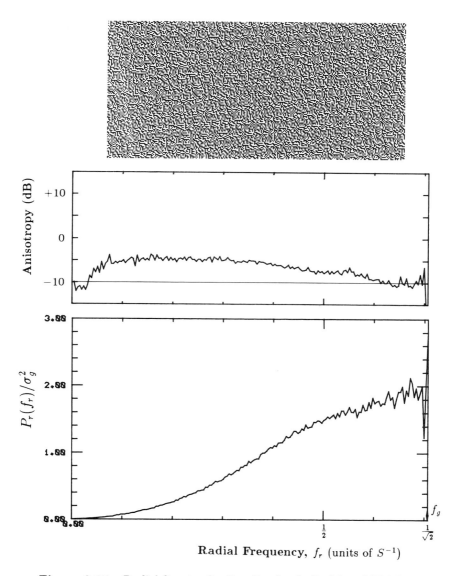

Figure 8.15: Radial Spectra for One Randomly Positioned Weight. (e) $\boxed{g = \frac{1}{2}}$, $f_g/S^{-1} \approx .7071$, $\sigma_g^2 = .25$

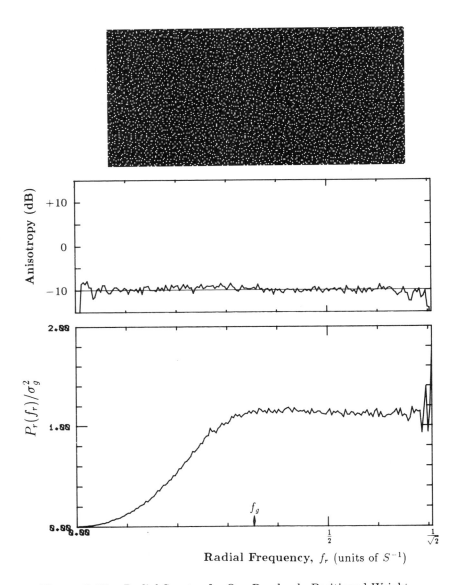

Figure 8.15:   Radial Spectra for One Randomly Positioned Weight. (f) $\boxed{g = \tfrac{7}{8}}$, $f_g/S^{-1} \approx .3536$, $\sigma_g^2 \approx .1094$

$$\left(\tfrac{1}{2}\times\right) \qquad \begin{matrix} \bullet & 1 \\ 1 & \end{matrix}$$

Figure 8.16: Deterministic Part of a Two Weight Error Filter.

8.3.2 Two Weights

In this section, four variations on the error filter shown in Figure 8.16 will be considered.

As might be expected, using this filter unperturbed yields unacceptable results as seen in Figure 8.17. The strong diagonal texture patterns result in an extremely anisotropic power spectrum evidenced by Figure 8.18.

The use of a serpentine raster corrects the directionality of these textures but still leaves many undesirable patterns. This is shown in Figure 8.19.

Figure 8.20 shows the result of adding the perturbation of 100% randomness to the two weights. This is a particularly interesting case because it simply requires the selection of a random number distributed between 0 and 1 for one weight, and its two's complement for the other.

The very random nature of this approach, while making the patterns more isotropic, passed too much low frequency content. A compromise is seen in Figure 8.21 where the weights were perturbed with 50% noise. The well behaved radially averaged power spectra results in Figure 8.22 (pages 288 through 292).

Figure 8.17: Effect of Two Deterministic Weights.

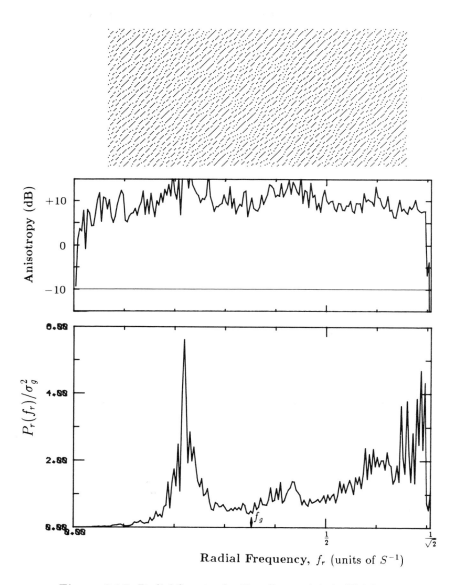

Figure 8.18: Radial Spectra for Two Deterministic Weights. $\boxed{g = \frac{1}{8}}$, $f_g/S^{-1} \approx .3495$, $\sigma_g^2 \approx .1094$

Figure 8.19: Effect of 2 Deterministic Weights on a Serpentine Raster.
(a) Gray Scale Ramp, $\alpha_r = 1$.

Figure 8.19: Effect of 2 Deterministic Weights on a Serpentine Raster. (b) Scanned Picture, $\alpha_r = 1$.

Figure 8.20: Effect of Two 100% Random Weights on a Serpentine Raster. (a) Gray Scale Ramp, $\alpha_r = 1$.

Figure 8.20: Effect of Two 100% Random Weights on a Serpentine Raster.
(b) Scanned Picture, $\alpha_r = 1$.

Figure 8.21: Effect of Two 50% Random Weights on a Serpentine Raster. (a) Gray Scale Ramp, $\alpha_r = 1$.

Figure 8.21: Effect of Two 50% Random Weights on a Serpentine Raster. (b) Scanned Picture, $\alpha_r = 1$.

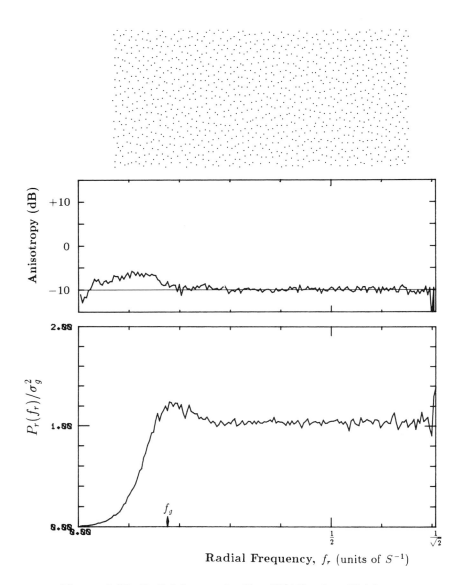

Figure 8.22: Radial Spectra for Two 50% Random Weights.

(a) $\boxed{g = \frac{1}{32}}$, $f_g/S^{-1} \approx .1768$, $\sigma_g^2 \approx .0303$

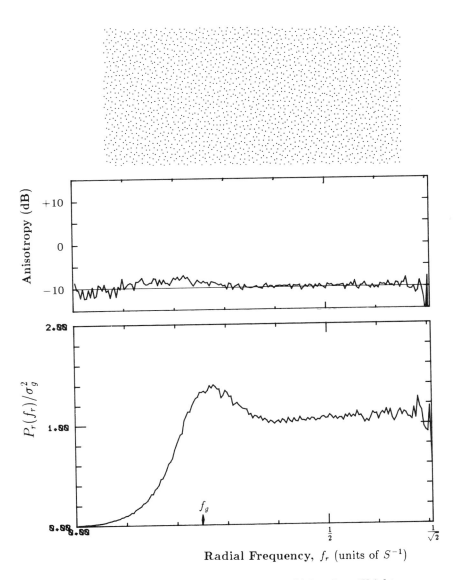

Figure 8.22: Radial Spectra for Two 50% Random Weights.

(b) $\boxed{g = \frac{1}{16}}$, $f_g/S^{-1} = .25$, $\sigma_g^2 \approx .0586$

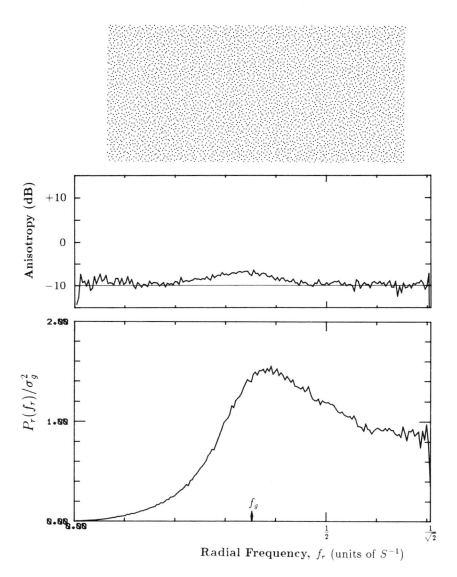

Figure 8.22:   Radial Spectra for Two 50% Random Weights.
(c) $\boxed{g = \frac{1}{8}}$, $f_g/S^{-1} \approx .3495$, $\sigma_g^2 \approx .1094$

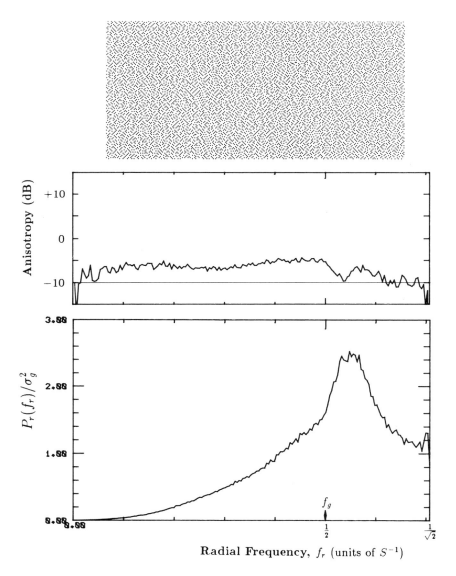

Figure 8.22: Radial Spectra for Two 50% Random Weights.
(d) $\boxed{g = \frac{1}{4}}$, $f_g/S^{-1} = .5$, $\sigma_g^2 = .1875$

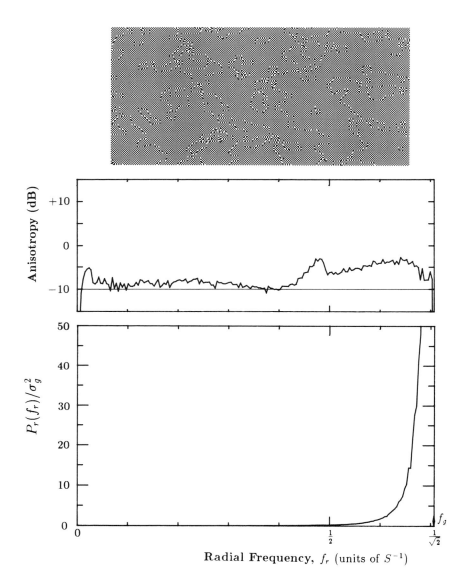

Figure 8.22: Radial Spectra for Two 50% Random Weights.
(e) $\boxed{g = \frac{1}{2}}$, $f_g/S^{-1} \approx .7071$, $\sigma_g^2 = .25$

8.3.3 Four Weights

Although better than that which can be achieved with one weight, the transition regions for the two weight case just examined were still not as steep as desired. Experiments with various choices of three weights did not produce a significant improvement.

In trying several combinations of deterministic values in a four element error filter, none proved better than the famed filter of Floyd and Steinberg. In this section, two variations on this basic filter are presented, both processed with serpentine rasters. Figure 8.23 shows the results of error diffusion holding the filter values constant but perturbing the threshold by 30% with white noise. The serpentine raster used in processing is responsible for much of the directional artifact elimination. The noisy threshold breaks up most remaining stable texture patterns yielding good radial symmetry at the expense of adding some low frequency energy.

In Figure 8.24, instead of perturbing the threshold, noise was added to the weights. For this purpose, the two larger weights ($\frac{7}{16}$ and $\frac{5}{16}$) and the two smaller weights ($\frac{3}{16}$ and $\frac{1}{16}$) were paired together. To prevent weights with negative values, the maximum noise amplitude (100%) is the value of the smaller weight in each pair. 50% noise was added to each pair in this case.

The radially averaged power spectra in Figure 8.25 (pages 299 through 305) reveal a good blue noise process. Along with low anisotropy and flat high frequency regions, the extra number of weights provided additional low frequency inhibition and steeper transition regions.

Here, as in many of the other cases considered, the most disturbing patterns evolve near $g = \frac{1}{2}$. Unless the power spectra contain impulses at the corners of the frequency baseband, perturbations to this perfect spatial checkerboard pattern, no matter how slight, become very visible. The fixed background in the synthesized image of Figure 8.24(c), which is near $g = \frac{1}{2}$, is an extreme example. Rendering gray levels near this value is perhaps the greatest weakness of the error diffusion algorithm and variations on it.

Figure 8.23: Floyd and Steinberg Filter with a 30% Random Threshold
Processed on a Serpentine Raster.
(a) Gray Scale Ramp, $\alpha_r = 1$.

Figure 8.23: Floyd and Steinberg Filter with a 30% Random Threshold
Processed on a Serpentine Raster.
(b) Scanned Picture, $\alpha_r = 1$.

Figure 8.24: Floyd and Steinberg Filter with 50% Random Weights
Processed on a Serpentine Raster.
(a) Gray Scale Ramp, $\alpha_r = 1$.

Figure 8.24: Floyd and Steinberg Filter with 50% Random Weights
Processed on a Serpentine Raster.
(b) Scanned Picture, $\alpha_r = 1$.

Figure 8.24: Floyd and Steinberg Filter with 50% Random Weights Processed on a Serpentine Raster. (c) Synthesized Image, $\alpha_r = 1$.

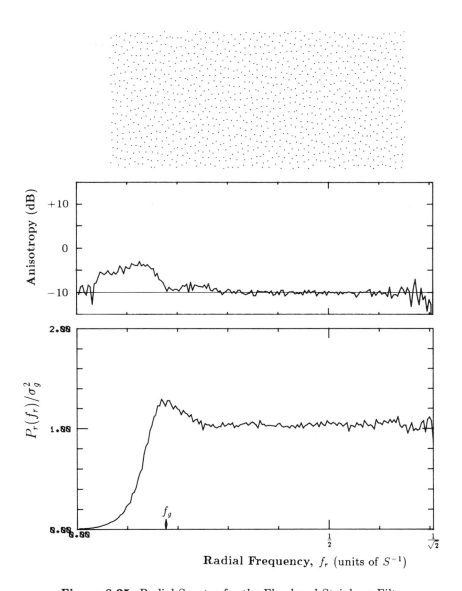

Figure 8.25: Radial Spectra for the Floyd and Steinberg Filter with 50% Random Weights Processed on a Serpentine Raster. (a) $\boxed{g = \frac{1}{32}}$, $f_g/S^{-1} \approx .1768$, $\sigma_g^2 \approx .0303$

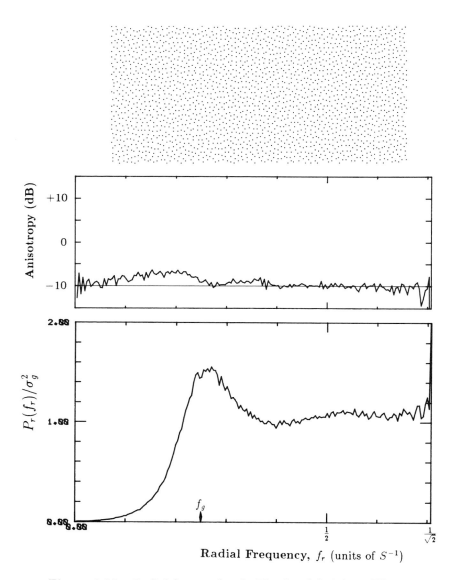

Figure 8.25:   Radial Spectra for the Floyd and Steinberg Filter
with 50% Random Weights Processed on a Serpentine Raster.
(b) $\boxed{g = \frac{1}{16}}$, $f_g/S^{-1} = .25$, $\sigma_g^2 \approx .0586$

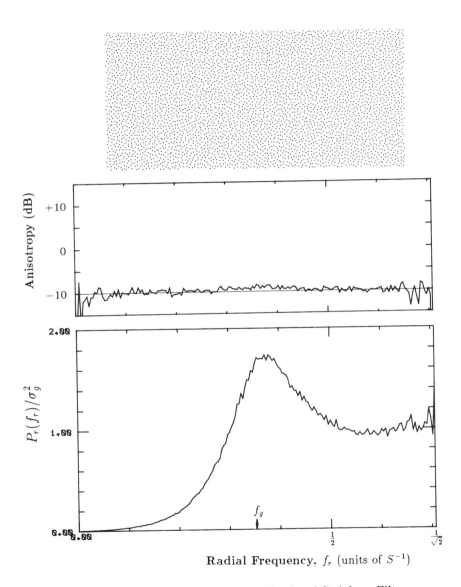

Figure 8.25: Radial Spectra for the Floyd and Steinberg Filter with 50% Random Weights Processed on a Serpentine Raster.

(c) $\boxed{g = \frac{1}{8}}$, $f_g/S^{-1} \approx .3495$, $\sigma_g^2 \approx .1094$

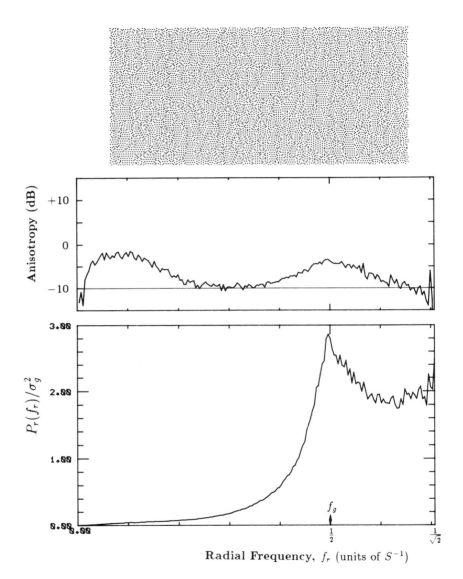

Figure 8.25: Radial Spectra for the Floyd and Steinberg Filter
with 50% Random Weights Processed on a Serpentine Raster.

(d) $\boxed{g = \frac{1}{4}}$, $f_g/S^{-1} = .5$, $\sigma_g^2 = .1875$

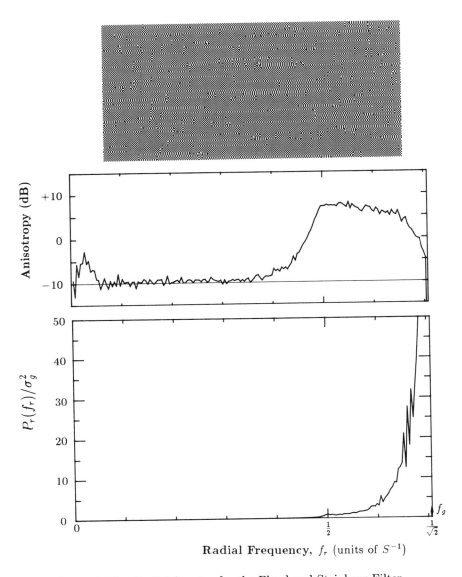

Figure 8.25: Radial Spectra for the Floyd and Steinberg Filter with 50% Random Weights Processed on a Serpentine Raster.

(e) $\boxed{g = \frac{1}{2}}$, $f_g/S^{-1} \approx .7071$, $\sigma_g^2 = .25$

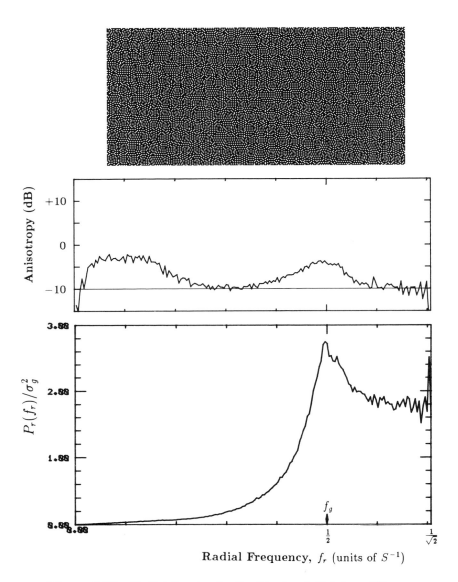

Figure 8.25: Radial Spectra for the Floyd and Steinberg Filter with 50% Random Weights Processed on a Serpentine Raster.

(f) $\boxed{g = \frac{3}{4}}$, $f_g/S^{-1} = .5$, $\sigma_g^2 = .1875$

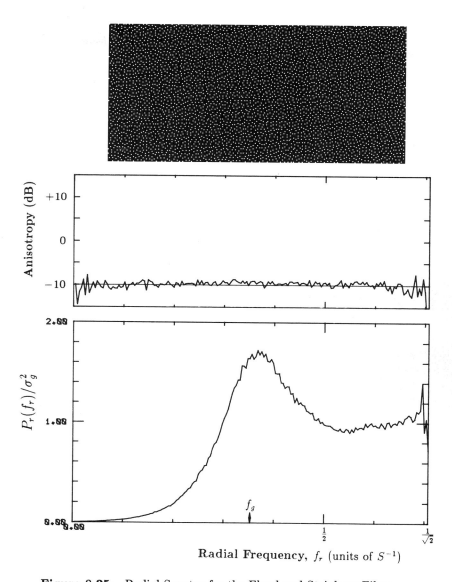

Figure 8.25: Radial Spectra for the Floyd and Steinberg Filter with 50% Random Weights Processed on a Serpentine Raster.

(g) $\boxed{g = \frac{7}{8}}$, $f_g/S^{-1} \approx .3536$, $\sigma_g^2 \approx .1094$

If plots of $P_r(f_r)$ at several other gray levels could be added to Figure 8.25 and shown in sequence, the form of the plot would look like a wave beginning at $f_r = 0$ at $g = 0$ moving to the right as g increased until it hit the "high frequency wall" at $g = \frac{1}{2}$. Then as g increased from $\frac{1}{2}$ to 1, the "wave" would retreat in a symmetric fashion.

8.3.4   Asymmetric Robustness

Figure 8.26 demonstrates the ability of error diffusion to perform reasonably well on grids with asymmetric aspect ratios. In the coarse example shown here, $\alpha_r = \frac{1}{6}$. This should be compared with the images of Chapter 7 with the same aspect ratio. As in the other asymmetric examples with $\alpha = \frac{1}{6}$, the resolution in pixels per unit area of the images shown is reduced by a factor of 6; the width of the pixels in the horizontal direction is fixed at the minimum that will survive reproduction.

While uncompensated ordered dither clearly fails at this aspect ratio, when properly compensated for $\alpha_r = \frac{1}{6}$, it suffers less anisotropicity than uncompensated error diffusion. The power spectrum of gray levels produced on asymmetric grids by error diffusion will certainly no longer be radially symmetric; however, because spectral energy is not concentrated at points as in the ordered dither case, the contortions of asymmetry are more diluted.

Even though error diffusion is somewhat robust in this regard, it could benefit from some compensation. As of yet, the nature of compensated error filters for asymmetry remains an open problem.

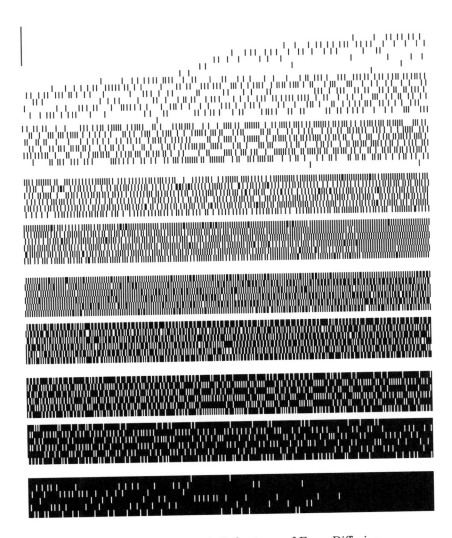

Figure 8.26: Asymmetric Robustness of Error Diffusion.
Floyd and Steinberg Filter with 50% Random Weights
Processed on a Serpentine Raster.
(a) Gray Scale Ramp, $\alpha_r = \frac{1}{6}$ (vertical resolution decrease).

Figure 8.26: Asymmetric Robustness of Error Diffusion.
Floyd and Steinberg Filter with 50% Random Weights
Processed on a Serpentine Raster.
(b) Scanned Picture, $\alpha_r = \frac{1}{6}$ (vertical resolution decrease).

8.4 Hexagonal Case

In section 8.1 it was argued that hexagonal grids were inferior to rectangular grids as far as generating blue noise. This is based on the inability of a hexagonal grid to support a principal wavelength, λ_g, for $\frac{1}{3} < g < \frac{2}{3}$ (see Figure 8.1(b), page 235). In spite of this deficiency, the other features of hexagonal grids, particularly its superior covering efficiency (Figure 2.5), are still reason to devote attention.

The only reported attempt at performing error diffusion on a hexagonal grid was by Stevenson and Arce [76], whose error filter was given Figure 8.6(d) on page 241. They stated that this is the filter which gave the "highest image quality", but admitted that no optimization was done. The effect of hexagonal error diffusion with this filter is shown in Figure 8.27. The gray scale ramp reveals many disturbing texture patterns.

A close look at the radially averaged power spectrum in Figure 8.28 reflects the large measure of anisotropy in 7 selected patterns. Perhaps the reason for the bizarre shape of many of the anisotropy plots is due to the stable texture patterns that begin to "grow" in regions of constant gray producing localized spikes in the power spectra.

Figure 8.27: Error Diffusion with the Stevenson and Arce Filter.
(a) Gray Scale Ramp, $\alpha_h = \frac{2}{\sqrt{3}}$.

Figure 8.27: Error Diffusion with the Stevenson and Arce Filter. (b) Scanned Picture, $\alpha_h = \frac{2}{\sqrt{3}}$.

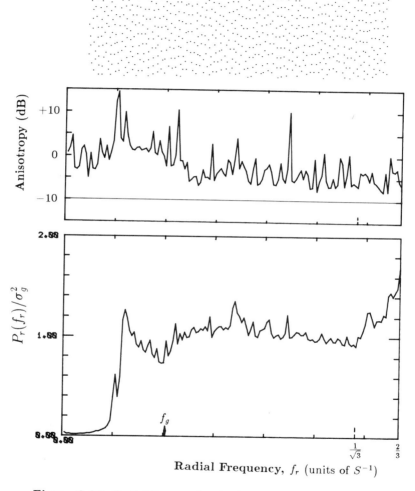

Figure 8.28: Radial Spectra for the Stevenson and Arce Filter.
(a) $\boxed{g = \frac{1}{32}}$, $f_g/S^{-1} \approx .2041$, $\sigma_g^2 \approx .0303$

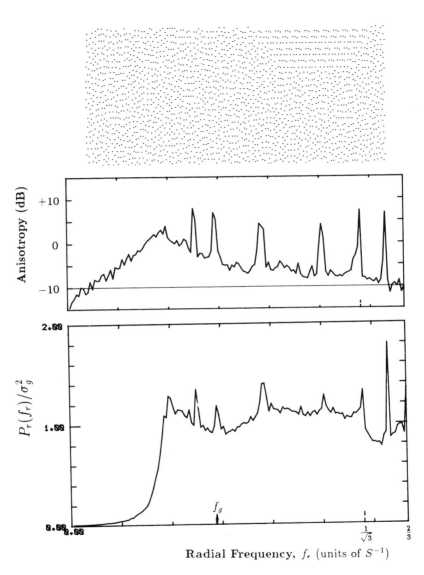

Figure 8.28: Radial Spectra for the Stevenson and Arce Filter. (b) $\boxed{g = \frac{1}{16}}$, $f_g/S^{-1} \approx .2887$, $\sigma_g^2 \approx .0586$

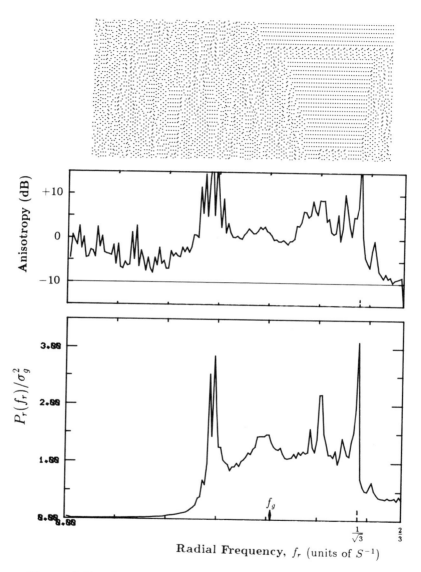

Radial Frequency, f_r (units of S^{-1})

Figure 8.28:   Radial Spectra for the Stevenson and Arce Filter.
(c) $\boxed{g = \frac{1}{8}}$, $f_g/S^{-1} \approx .4082$, $\sigma_g^2 \approx .1094$

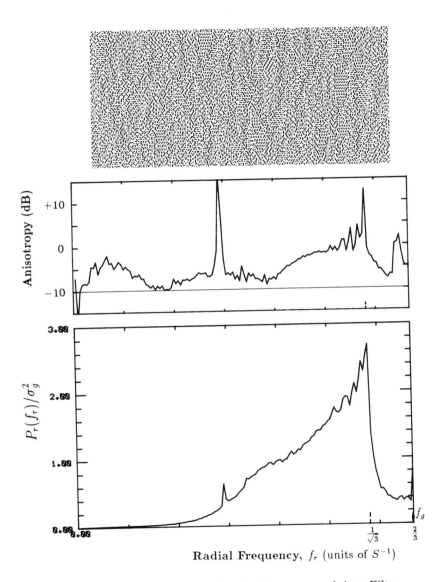

Figure 8.28: Radial Spectra for the Stevenson and Arce Filter.
(d) $\boxed{g = \frac{1}{3}}$, $f_g/S^{-1} = \frac{2}{3}$, $\sigma_g^2 \approx .1111$

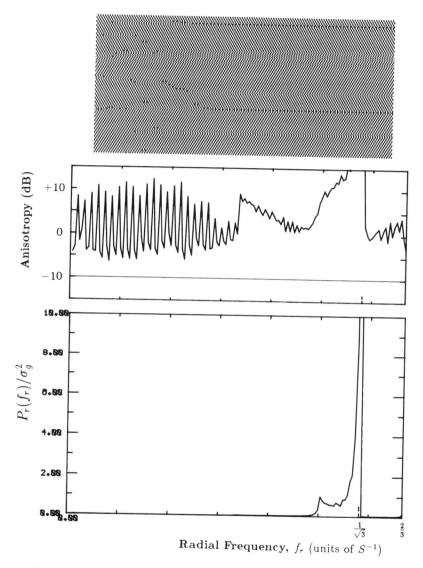

Figure 8.28: Radial Spectra for the Stevenson and Arce Filter.

(e) $\boxed{g = \frac{1}{2}}$, (f_g is not supported), $\sigma_g^2 = .25$

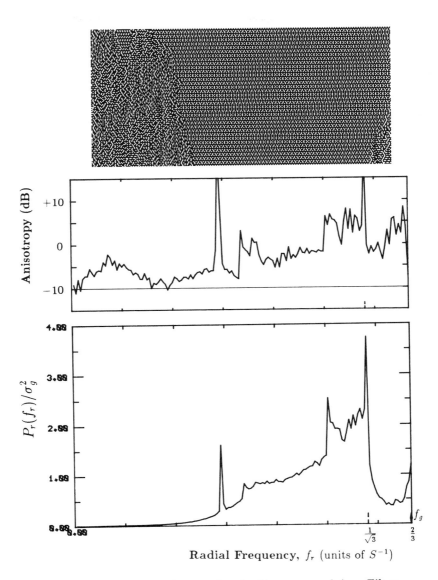

Figure 8.28: Radial Spectra for the Stevenson and Arce Filter.
(f) $\boxed{g = \frac{2}{3}}$, $f_g/S^{-1} = \frac{2}{3}$, $\sigma_g^2 = .1111$

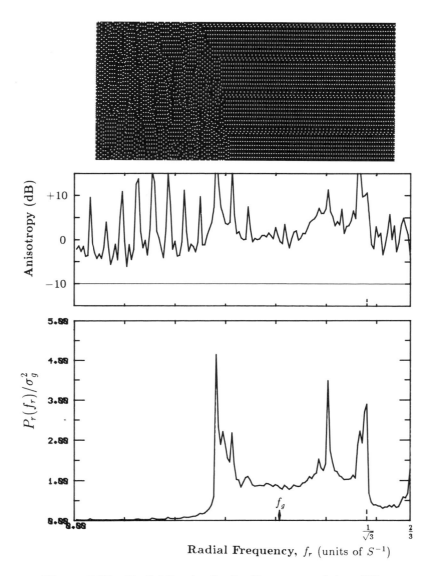

Figure 8.28: Radial Spectra for the Stevenson and Arce Filter.
(g) $\boxed{g = \frac{7}{8}}$, $f_g/S^{-1} \approx .4082$, $\sigma_g^2 \approx .1094$

$$\left(\frac{1}{2}\times\right) \qquad\qquad \begin{matrix}\bullet & 1 \\ 1 & \end{matrix}$$

(a) Two Weights.

$$\left(\frac{1}{16}\times\right) \qquad\qquad \begin{matrix} & \bullet & 7 \\ 3 & 5 & 1\end{matrix}$$

(b) Variant of Floyd and Steinberg Filter.

Figure 8.29: Two Hexagonal Error Filters to be Perturbed.
("•" represents the origin.)

8.4.1   Error Diffusion with Perturbation

It turns out that the most successful methods of halftoning by error diffusion on rectangular grids works quite well on hexagonal grids after adjusting the error filters slightly. Figure 8.29 displays the deterministic part of the two and four element filter that will be demonstrated in this section.

Parallel to the methods used for Figures 8.21 and 8.24, these filters will have 50% noise added to their weights, and processing will be on a serpentine raster. The result of error diffusion with the two element filter is illustrated in Figure 8.30; the result of the four element filter is shown in Figure 8.31.

Both render gray levels in a much more isotropic manner than that of Figure 8.27. The large 12 element filter did sharpen the picture more, however. As in the rectangular case, because of the asymmetric nature and lack of control, it is better if sharpening is not compounded with halftoning. In the next chapter, an example is given of how precisely controlled sharpening can be achieved in a separate operation for both rectangular and hexagonal grids.

Figure 8.30: Effect of Two 50% Random Weights on a Hexagonal Grid
Processed on a Serpentine Raster.
(a) Gray Scale Ramp, $\alpha_h = \frac{2}{\sqrt{3}}$.

Figure 8.30: Effect of Two 50% Random Weights on a Hexagonal Grid Processed on a Serpentine Raster.
(b) Scanned Picture, $\alpha_h = \frac{2}{\sqrt{3}}$.

Figure 8.31: Effect of Four 50% Random Weights on a Hexagonal Grid
Processed on a Serpentine Raster.
(a) Gray Scale Ramp, $\alpha_h = \frac{2}{\sqrt{3}}$.

Figure 8.31: Effect of Four 50% Random Weights on a Hexagonal Grid
Processed on a Serpentine Raster.
(b) Scanned Picture, $\alpha_h = \frac{2}{\sqrt{3}}$.

The radially averaged power spectrum of both the two and four weight stochastic hexagonal filters are well behaved with the four weight case having sharper transition regions. So this case is included in Figure 8.32. Notice how well the peak of these spectra follow the principal frequency (marked as usual with a diamond on the frequency axis). The important exception is for $g = \frac{1}{2}$ which is in the region where the hexagonal grid cannot support a principal frequency.

Two other particularly interesting cases are those for $g = \frac{1}{3}$ and $g = \frac{2}{3}$. The principal frequency for these gray values is at the high frequency limit for a hexagonal grid, $f_g = \frac{2}{3}$. These cases are similar to the $g = \frac{1}{2}$ case for rectangular grids.

In balance, anisotropy is generally low for the stochastic hexagonal filter, not nearly as wild as in Figure 8.28. Outside of the range $\frac{1}{3} < g < \frac{2}{3}$, such filters are also good blue noise generators.

The wave analogy given for rectangular grid blue noise spectra can also be used for hexagonal grids. A sequence of plots of $P_r(f_r)$ for several gray levels would show a "wave" beginning at $f_r = 0$ at $g = 0$ moving to the right as g increased until it hit the "high frequency wall" at $g = \frac{1}{3}$. Skipping to $g = \frac{2}{3}$, the "wave" would retreat in a symmetric fashion as g increased from $\frac{2}{3}$ to 1.

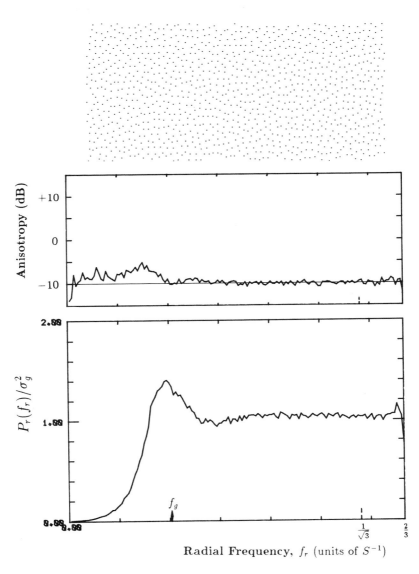

Figure 8.32: Radial Spectra for the Stochastic Hexagonal Filter. (a) $\boxed{g = \frac{1}{32}}$, $f_g/S^{-1} \approx .2041$, $\sigma_g^2 \approx .0303$

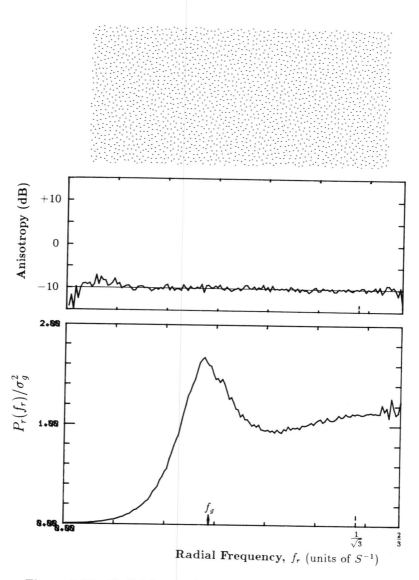

Figure 8.32:   Radial Spectra for the Stochastic Hexagonal Filter.
(b) $\boxed{g = \frac{1}{16}}$, $f_g/S^{-1} \approx .2887$, $\sigma_g^2 \approx .0586$

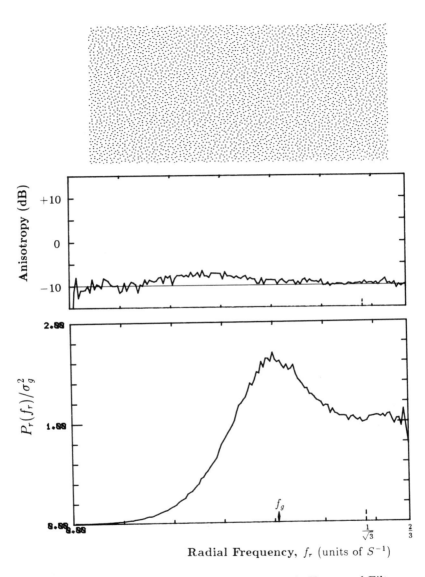

Figure 8.32: Radial Spectra for the Stochastic Hexagonal Filter.
(c) $\boxed{g = \frac{1}{8}}$, $f_g/S^{-1} \approx .4082$, $\sigma_g^2 \approx .1094$

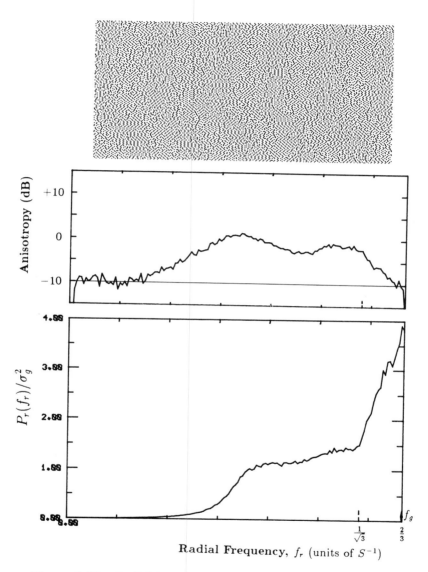

Figure 8.32: Radial Spectra for the Stochastic Hexagonal Filter.
(d) $\boxed{g = \frac{1}{3}}$, $f_g/S^{-1} = \frac{2}{3}$, $\sigma_g^2 \approx .1111$

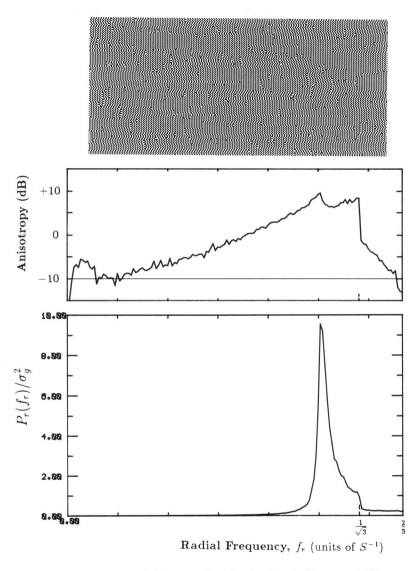

Figure 8.32: Radial Spectra for the Stochastic Hexagonal Filter.

(e) $\boxed{g = \frac{1}{2}}$, (f_g is not supported), $\sigma_g^2 = .25$

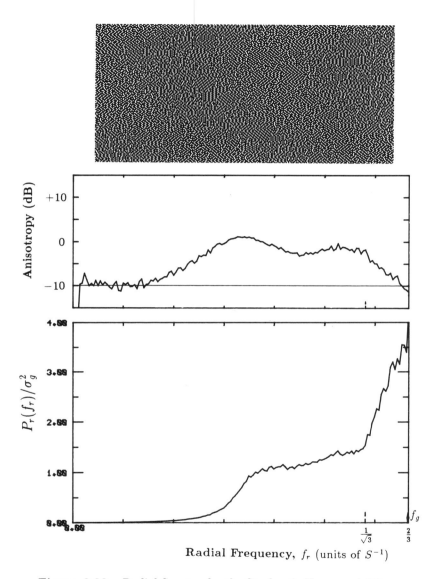

Figure 8.32:   Radial Spectra for the Stochastic Hexagonal Filter.
(f) $\boxed{g = \frac{2}{3}}$, $f_g/S^{-1} = \frac{2}{3}$, $\sigma_g^2 = .1111$

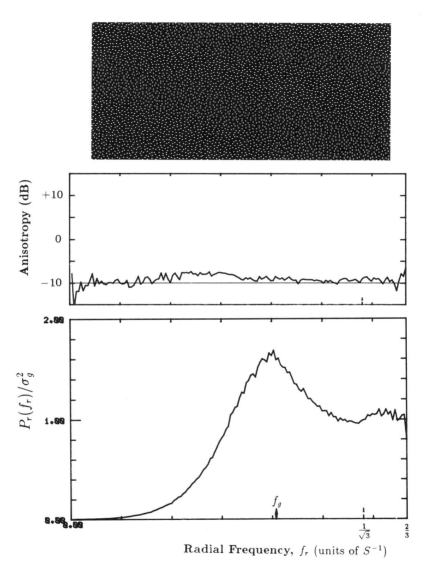

Figure 8.32: Radial Spectra for the Stochastic Hexagonal Filter.
(g) $\boxed{g = \frac{7}{8}}$, $f_g/S^{-1} \approx .4082$, $\sigma_g^2 \approx .1094$

Chapter 9

Concluding Remarks

For completeness, this chapter reviews some of the other halftoning related neighborhood process. In particular, the effect of presharpening an image with a digital Laplacian is demonstrated for both rectangular and hexagonal grids.

Patterns from each of the major classes of halftoning are compared, with blue noise found to exhibit a "grid defiance" property (to be described). The details on extrapolating halftone methods for use on multibit displays are explained.

At the end of this chapter, a summary of the major points of this text is provided with recommendations.

9.1   Other Neighborhood Processes

Besides the popular method of error diffusion, other neighborhood processes have been presented that use a local average to modify the halftone process. Jarvis's "Constrained Average" [35] algorithm is one such scheme. The success of the algorithm depends heavily on the digitization noise inherent in the picture. The threshold at any point is a function of the local average, dependent on the probability density function of the noise. If the noise level is too low, artificial "noise" must be added. Lippel [45] later complained that this algorithm was nothing more than a standard edge enhancing technique and that the value of its halftoning ability depended on the nature of the added noise.

Roetling [60] modified the classical screen by modulating the amplitude of the screen as a function of the local average; he called the algorithm ARIES, for Alias Reducing Image Enhancing Screen.

White [87] of IBM suggested separating the low and high frequencies and processing each separately. The low frequencies are represented by a clustered dot within an 8 pixel cell, which allows only 9 gray levels. A 3 by 3 Laplacian operator generates the high frequencies. An improvement on this idea was reported [7] that used a "nonlinear Laplacian" to reduce moire patterns when screening scanned classical halftones, and a larger (18 pixel) cell to better represent the low frequencies with less contouring.

9.1.1 Sharpening

As explained in section 1.2, the virtues of a halftoning scheme should be decoupled from its ability to sharpen. The improved output perceived from a method that intrinsically sharpens can misleadingly outweigh other shortcomings in its ability to render gray levels accurately and without algorithmic artifacts.

Sharpening does improve, or at least defeats, unsharpening degradations that halftoning imparts. The proper degree of sharpening is a subjective quality and can easily be controlled independently of halftoning. Sharpening can be combined with interpolation, that is, in the retrospective resampler of of the image rendering system of Figure 1.3 (page 8) for the case of digital enlarging. The sharpened Gaussian [69,70] is often used for such a dual purpose.

When sharpening is not combined with resampling, a separate high pass filter operation is needed. Perhaps the most popular high pass filter used for the purpose of sharpening is the digital Laplacian. The Laplacian, ∇^2, is an operator which produces the second spatial derivative,

$$\nabla^2 J(\mathbf{x}) = \frac{\partial^2 J(\mathbf{x})}{\partial x_1^2} + \frac{\partial^2 J(\mathbf{x})}{\partial x_2^2}. \tag{9.1}$$

When applied to an image, it produces large amplitudes at edge locations, and zero in constant or uniformly varying regions (regions where the zeroth or first derivative is zero).

This operator can be described as convolution with the filter $\nabla^2 \delta(\mathbf{x})$. The discrete-space analog to this on a rectangular grid is the 5 element

$$
\begin{array}{ccc}
 & \tfrac{1}{4} & \\
\tfrac{1}{4} & -1 & \tfrac{1}{4} \\
 & \tfrac{1}{4} &
\end{array}
\qquad \text{or} \qquad
\begin{array}{ccc}
\tfrac{1}{8} & \tfrac{1}{8} & \tfrac{1}{8} \\
\tfrac{1}{8} & -1 & \tfrac{1}{8} \\
\tfrac{1}{8} & \tfrac{1}{8} & \tfrac{1}{8}
\end{array}
$$

(a) Rectangular.

$$
\begin{array}{cccc}
 & \tfrac{1}{6} & \tfrac{1}{6} & \\
\tfrac{1}{6} & & -1 & & \tfrac{1}{6} \\
 & \tfrac{1}{6} & \tfrac{1}{6} &
\end{array}
$$

(b) Hexagonal.

Figure 9.1: Digital "Laplacian" Filters, $\Psi[\mathbf{n}]$.

filter shown in Figure 9.1(a), although the 9 element variety achieves a similar effect. A hexagonal version is shown in Figure 9.1(b).

Denoting a digital Laplacian filter as $\Psi[\mathbf{n}]$, sharpening is achieved by subtracting the Laplacian filtered image from the original image,

$$
J_{\text{sharp}}[\mathbf{n}] = J[\mathbf{n}] - \beta \Psi[\mathbf{n}] * J[\mathbf{n}]. \tag{9.2}
$$

The amount of sharpening is controlled by the value of $\beta \geq 0$. A tradeoff must be made between the accentuation of edge detail and amplification of noise.

An example of the effect of presharpening in this way on a rectangular grid with $\beta = 2.0$ is displayed in Figure 9.2. It should be compared to that on page 297 which was identically halftoned without presharpening. The hexagonal example in Figure 9.3 is the prefiltered version of that shown on page 168.

Adaptive sharpening techniques exist which are not as sensitive to noise but are, as one might expect, more compute intensive.

Figure 9.2: Sharpening with a Rectangular Laplacian Filter.
Laplacian amplitude, $\beta = 2.0$.
Halftone by error diffusion. Compare Figure 8.24(b) (page 297).

Figure 9.3: Sharpening with a Hexagonal Laplacian Filter.
Laplacian amplitude, $\beta = 2.0$.
Halftone by ordered dither. Compare Figure 6.13(e) (page 168).

9.2   Blue Noise is Pleasant

Figure 9.4 collectively compares greatly enlarged portions of the four major classes of dither patterns arranged in order of increasing correlation (decreasing entropy). All patterns are representations of a fixed gray level, $g = \frac{1}{9}$, and thus all have roughly the same number of black pixels.

While white noise appears too random or "noisy", ordered dither appears too "structured". The purpose of a dither pattern is to represent a continuous-tone level. It therefor should not have any form or structure of its own; a pattern succeeds when it is innocuous. Blue noise is visually pleasant because it does not clash with the structure of an image by adding one of its own or degrade it by being too "noisy" or uncorrelated.

Blue noise even defies the structure of the underlying grid. Even though the dots in Figure 9.4(b) are perfect squares, each precisely aligned to a given position on a rectangular grid, the collective ensemble tends to destroy this rigorous alignment creating what can be called a *grid defiance illusion*.

For many years, noise with $1/f$ power spectrum distributions have been known to exist in electrical systems. But recently, remarkable discoveries have repeatedly confirmed the existence of a $1/f$ power spectrum in almost every aspect of nature [25,47,86] including such things as variations in sunspots, wobbling of the earths axis, and flood levels of the River Nile. Evidence of $1/f$ fluctuations in human biological systems [52] has also been found; a $1/f$ spectrum was found in electroencephalogram (brain wave) measurements when subjects were exposed to "pleasing" stimuli.

A study by Richard Voss [84,85] has found that practically all forms of music possess $1/f$ noise. Experiments with stochastic music composition revealed that listeners found $1/f$ music far more interesting than white $(1/f^0)$ music, described as "too random", or brown $(1/f^2)$ music, described as "too correlated".

Blue noise can be described as the "pleasing" complement of $1/f$ noise. The dominance of low frequencies in $1/f$ phenomenon is responsible for its interesting and natural structure. Blue noise, by contrast,

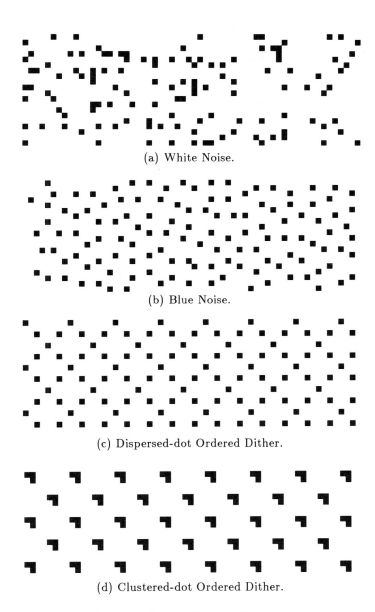

(a) White Noise.

(b) Blue Noise.

(c) Dispersed-dot Ordered Dither.

(d) Clustered-dot Ordered Dither.

Figure 9.4: Comparison of Halftone Patterns for $g = \frac{1}{9}$.
Arranged in order of decreasing entropy.

is not "interesting"; nor is it annoying. Being devoid of low frequencies and localized concentrations of spikes in the frequency domain, it has no structure and thus does not interfere with the interesting features of that which it is representing.

9.3   Beyond Binary Displays

As stated in the Introduction, halftoning or dithering can be used to redistribute coarse quantization noise whenever the number of gray levels to be displayed exceeds that which can be accommodated by the display device. The case of binary displays has been the focus of this book. The results are easily extrapolated to m-ary or multibit displays.

For systems of more than one bit, dither is not as critical as in binary displays. Thus not much attention has been paid to it. The old idea of random dither [26,58] works fairly well and is widely used in practice. Nonetheless, just as dithering with white noise proved to be inferior to other halftoning techniques in the binary case, dithering m-ary displays can also be improved.

For ordered dither, the threshold value was assumed to vary from 0 to 1 with the relative magnitudes governed by some threshold array. This process could be equivalently described by the relation,

$$I[\mathbf{n}] = \text{int}\{J[\mathbf{n}] + D[\mathbf{n}]\} \qquad (9.3)$$

where $D[\mathbf{n}]$ is a *dither signal* comprised of the ordered dither threshold values, $\text{int}\{\}$ represents integer truncation, and as usual, $J[\mathbf{n}]$ is the continuous-tone image, and $I[\mathbf{n}]$ is the halftoned image equal to either 0 or 1.

If the display could render 4 true levels (2 bits), dithering could be expressed as

$$I_{(2)}[\mathbf{n}] = \frac{1}{3}\text{int}\{3J[\mathbf{n}] + D[\mathbf{n}]\} \qquad (9.4)$$

where the quantized output, $I_{(2)}[\mathbf{n}] \in [0, \frac{1}{3}, \frac{2}{3}, 1]$. In general, ordered dither for a K bit display is

$$I_{(K)}[\mathbf{n}] = \frac{1}{2^K - 1}\text{int}\{(2^K - 1)J[\mathbf{n}] + D[\mathbf{n}]\}. \qquad (9.5)$$

This method assures that an input image, $J[\mathbf{n}]$, with a uniformly distributed histogram will result with the same discrete histogram as quantizing with fixed thresholds.

The output levels of a fixed K-bit quantizer would included 0 and 1 with the remaining 2^{K-1} levels equispaced between them. The $2^K - 1$ threshold levels are equispaced between the output values. Such a quantizer would replace the fixed threshold of a K-bit error diffusion algorithm. The other aspects of the algorithm, along with all of the permutations of Chapter 8 remain unaltered. It is interesting to note that as early as 1969, an m-ary error distribution algorithm similar to error diffusion was demonstrated on a multiple exposure microfilm recorder [71].

The most efficient way to distribute a finite set of gray levels in a m-ary display device is *not* uniformly in reflectance or luminance, but in uniform steps of perceived brightness. (The relationship between brightness and luminance vary nonlinearly in deferent ways for video and reflection copy). If the Physical Reconstruction Function of the device has such properly separated gray levels, the compensation should be handled in the Tone Scale Adjust portion of the display preprocessor (Figure 1.3, page 8) and not in the quantizer.

An important open issue concerning m-ary displays is the tradeoff in terms of image quality between spatial and amplitude resolution, holding the total number of image bits constant.

9.3..1 Color

These same dithering techniques are also easily extended to multispectral displays. In general, quantizing and displaying color images while faithfully preserving chromaticity is a complex and delicate problem. The fundamentals of color hard copy are presented by Yule [90]; Engeldrum [19] discusses the variables that specifically contribute to color gamuts of dot formed printing systems.

In many video display systems, the frame buffer is limited to K bits per pixel, but each of the 2^K pixel values can be mapped to any of a much larger set of colors by means of a look-up table. So there is a choice of which 2^K colors to use. Heckbert [30] reviews color quantization issues, and suggests choosing a color map based on the color statistics of the particular image to be rendered. He also showed that error diffusion can be used to mitigate quantization noise by demonstrating fairly good results with only two bits per pixel.

Color error diffusion can be considered a three dimensional extension of the m-ary case described above. The quantizer (which can be designed using Heckbert's method) can be thought of as 2^K output points in some 3-D color space, separated by threshold *surfaces*. The "error" to be distributed to future input values would be the *vector* between the input value and the quantized output value. An important design consideration is the color space in which the quantization occurs, so that the error vector or distance measure is perceptually uniform over that space.

Practical hardcopy printers, unlike video systems, lack the luxury of selecting an arbitrary set of output colors; typically three or four components which can be overlaid are available. By far, the most studied case of color halftoning is that of four color printing with the classical screen as discussed in section 5.2.2.

For dispersed-dot ordered dither, the common practice in hardcopy is to simply halftone each component color independently and overprint. A common application involves dithering each component of a source image coded with the additive primaries—red, green, and blue—resulting in the 3 bits $(b_R b_G b_B)$. A color hard copy device can typically display the three subtractive primaries—cyan, magenta, and yellow—with each pixel represented by the bits $(b_C b_M b_Y)$. The simple transformation

$$
\begin{aligned}
b_C &= \bar{b}_R \\
b_M &= \bar{b}_G \\
b_Y &= \bar{b}_B
\end{aligned}
$$

where $\bar{b}$ refers to the logical complement of b, is used to produce the data to be printed. The problem of morié patterns due to beat frequencies between overprinted patterns is not appreciable in the case or dispersed-dot ordered dither.

9.4 Summary

A comprehensive investigation of new and old techniques available for digital halftoning has been presented. The corresponding digitally produced examples can serve as a catalog for evaluating the tradeoffs between methods.

Halftoning is but one component in a total image rendering system. The production of high quality images requires a preprocessor tailored to the peculiarities of a target Physical Reconstruction Function. Before halftoning, an image must be appropriately resampled, tone scale adjusted and optionally sharpened.

A clustered-dot technique should only be used if the Physical Reconstruction Function (which might include a reproduction process) cannot accommodate isolated black or white pixels. Otherwise, a dispersed-dot method should always be used.

In this text nonstandard grid geometries were considered. The "aspect ratio immunity" argument revealed that semiregular hexagonal grids outperform the *best* covering efficiency of rectangular grids over an order of magnitude of aspect ratio. Dot overlap decreases with improved covering efficiency, so tone scale can be more precisely controlled on hexagonal grids.

The advantages of hexagonal displays can be realized in most cases without converting all storage formats in an imaging system to hexagonal form; some classes of images such as text are better rendered on rectangular grids [57]. For the case where a rectangularly sampled continuous-tone image is to be produced on binary displays of lower resolution, subsampling could be performed in a hexagonal fashion before halftoning. As was pointed out earlier, the hardware modification needed to convert a rectangular display to a semiregular hexagonal one is quite simple; an alternate line offset of one half sample period is all that is required.

The point process of ordered dither was generalized for both rectangular and hexagonal grids by the Method of Recursive Tessellation. In the case of asymmetric grids, the "immunity" of hexagonal grids can be used even in the case of rectangular grids. A set of guidelines defining the number of ternary replications required in the generation of asymmetric threshold arrays to create patterns with maximum symmetry as a function of aspect ratio has been presented.

The concept of dithering with blue noise along with its "grid defiance" property was introduced. Conventional methods of error diffusion with previously reported error filters were closely examined and found to be fair blue noise generators. Experiments with a broad array of perturbations found that excellent blue noise patterns could be achieved with error filters of four or fewer weights when noise was added and processed on a serpentine raster. While more computationally expensive than ordered dither, a blue noise technique should be used when ever the resolution of a display device is so low that every pixel is clearly visible. Being devoid of low frequencies, it does not interfere with the structure of the image it is rendering.

However, dithering with blue noise on hexagonal grids was inferior to rectangular grids as predicted from theoretical arguments in the frequency domain. This is perhaps a surprising result, considering the success of ordered dither on such grids.

The nature of the patterns resulting from halftoning uniform planes of constant gray level were evaluated in the frequency domain for rectangular and hexagonal grids. New metrics for analyzing the frequency content of both periodic and aperiodic patterns were developed. For the periodic patterns resulting from ordered dither, insight was gained by displaying composite Fourier transforms in the form of "exposure plots". Radially average power spectra combined with a measure of anisotropy were used to evaluate aperiodic patterns.

This collection of halftoning schemes should prove useful in the implementation of image networks that employ a wide variety of practical displays.

Glossary of Principal Symbols

The following notational conventions are followed in this book:

- Bold lower case symbols are vectors.

- Bold upper case symbols are matrices.

- All other (script) symbols are scalars.

- Arguments of discrete-space signals are enclosed in square brackets, [ ].

- Arguments of continuous-space signals are enclosed in parentheses, ( ).

This alphabetically arranged glossary presents Greek symbols first, followed by English symbols. The italicized number following each description is the page number in this text where the symbol is defined or illustrated.

α	Aspect ratio. It equals the Sample Period divided by the Line Period, S/L. *(20)*
β	Magnitude of a Laplacian to be subtracted from an image for the purpose of sharpening. It is proportional to the degree of sharpening. *(335)*
$\delta()$	Impulse function. Has the property of being equal to zero everywhere except where the argument is zero, and has

unit "area", $\int_{-\infty}^{\infty} \delta(\mathbf{x})d\mathbf{x} = 1$. Bracewell [12, ch. 5] has a good discussion of this function.

Δ Width of the annuli over which the power spectrum estimate, $\hat{P}(\mathbf{f})$, is averaged to form the radially averaged power spectrum, $P_r(f_r)$.   *(56,57)*

η Order or size of an ordered dither threshold matrix which was generated by the method of recursive tessellation. Rectangular and hexagonal matrices have 2^η and 3^η elements, respectively.   *(128)*

κ Number of ternary subdivisions and replications necessary to compensate an ordered dither threshold array for use on an asymmetric grid.   *(196)*

λ_g Principal radial wavelength in a homogeneously distributed field of binary pixels representing the constant gray level, g.   *(234)*

σ_g^2 Variance of an individual output binary pixel resulting from halftoning homogeneous region of gray level g.   *(53,59)*

$\Psi[\mathbf{n}]$ Digital Laplacian filter used for presharpening an image prior to halftoning.   *(335)*

$b(\mathbf{x})$ Global Background. The linear but shift-varying global background luminance contribution in the Physical Reconstruction Function.   *(16,17)*

$c(\mathbf{x})$ Continuous-space image. Output of the Discrete-to-Continuous Space Converter.   *(16,17,44)*

$C(\mathbf{f})$ Continuous-space Fourier Transform of image $c(\mathbf{x})$. *(45)*

$C_\Sigma(\mathbf{f})$ Composite Fourier Transform. The average of the Fourier transform magnitudes of all possible gray level renderings produced by thresholding with a particular ordered dither matrix.   *(53)*

$d(\mathbf{x})$ Dot function. The linear, shift invariant function which models the visible output pixel in the Physical Reconstruction Function.   *(16,27)*

$e[\mathbf{n}]$ Error Filter. Governs how past quantization errors are negatively distributed or "diffused" into the yet to be quantized image in the error diffusion algorithm.   *(240)*

$E(\alpha)$ Efficiency of a periodic sampling grid as a function of aspect ratio. Defined as the ratio of pixel area (see pages 19 and 18 for definition of pixel shape) to the area of a circumscribing circle.   *(24,25)*

$\mathbf{f}$ Continuous-space frequency vector, $\mathbf{f} = \begin{bmatrix} f_1 \\ f_2 \end{bmatrix}$.   *(45)*

f_g Principal radial frequency in a field of homogeneously distributed binary pixels representing gray level g.   *(234)*

f_r Radial frequency. The scalar distance in frequency units from zero frequency in a two dimensional Fourier transform.   *(56,57)*

g Gray level. It has a continuous value in the range 0 (white) to 1 (black) inclusive.   *(35)*

$I[\mathbf{n}]$ Quantized discrete-space image. The output from a halftoning process.   *(8,6,16)*

$I[\mathbf{n}; g]$ The binary output image resulting from halftoning an image consisting of a fixed gray level, $J[\mathbf{n}] = g$.   *(52)*

$I[\mathbf{k}]$ Discrete Fourier Transform (DFT) of $I[\mathbf{n}]$.   *(46)*

$I[\mathbf{k}; g]$ The DFT of the binary output image resulting from halftoning an image consisting of a fixed gray level, $J[\mathbf{n}] = g$.   *(52)*

$I_\Sigma[\mathbf{k}]$ Composite DFT. The average of the Discrete Fourier Transform magnitudes of all possible gray level renderings produced by thresholding with a particular ordered dither matrix, that is, the average of $I[\mathbf{k}; g]$ for all g.   *(52)*

$I(\mathbf{x})$ Visible Image. Output of the Physical Reconstruction Function.   *(8,6)*

$J[\mathbf{n}]$ Continuous amplitude, discrete-space image. Input to a halftoning process.   *(8,9,16)*

$\mathbf{k}$ Discrete-space frequency index vector, $\mathbf{k} = \begin{bmatrix} k_1 \\ k_2 \end{bmatrix}$.

k_1 and k_2 are integers.   *(46)*

L Line period.   *(18,19)*

$\mathcal{L}$ Luminance. That photometric quantity which is perceived, having units of lumens/area/steradian.   *(35)*

$\mathbf{n}$ Discrete-space spatial index vector, $\mathbf{n} = \begin{bmatrix} n_1 \\ n_2 \end{bmatrix}$.

n_1 and n_2 are integers.

$N_r(f_r)$ Number of discrete frequency samples in an annulus about radial frequency f_r.   *(56,61)*

$\mathbf{p}_1, \mathbf{p}_2$ Spatial period replication vectors.   *(42,50)*

$\mathbf{P}$ Spatial period replication matrix, $[\mathbf{p}_1 \vdots \mathbf{p}_2]$.   *(46)*

$P(\mathbf{f})$ Power Spectrum. In this book, only the power spectra of binary output of aperiodic halftone processes on a single input gray level are considered.   *(54)*

$\hat{P}(\mathbf{f})$ Power Spectrum Estimate.   *(54)*

$P_r(f_r)$ Radially Averaged Power Spectrum. Sample mean of the frequency samples of $\hat{P}(\mathbf{f})$ in the annulus, $||\mathbf{f}| - f_r| < \Delta/2$, about f_r.   *(56)*

$\mathbf{q}_1, \mathbf{q}_2$ Frequency sampling vectors.   *(46,50)*

$\mathbf{Q}$ Frequency sampling matrix, $[\mathbf{q}_1 \vdots \mathbf{q}_2]$.   *(46)*

R Reflectance. The ratio of reflected to incident radiant power.   *(35)*

$s^2(f_r)$ Sample variance of the frequency samples of $\hat{P}(\mathbf{f})$ in the annulus, $||\mathbf{f}| - f_r| < \Delta/2$, about f_r.   *(56)*

$s^2(f_r)/P_r^2(f_r)$ Anisotropy of $\hat{P}(\mathbf{f})$.   *(56)*

S Sample period.   *(18,19)*

$\mathbf{u}_1, \mathbf{u}_2$ Frequency baseband replication vectors.   *(45,50)*

$\mathbf{U}$ Frequency baseband replication matrix, $[\mathbf{u}_1 \vdots \mathbf{u}_2]$.   *(45)*

$\mathbf{v}_1, \mathbf{v}_2$ Spatial sampling vectors.   *(18,50)*

$\mathbf{V}$ Spatial sampling matrix, $[\mathbf{v}_1 \vdots \mathbf{v}_2]$.   *(18)*

$\mathrm{var}\{\hat{P}(\mathbf{f})\}$ Ensemble variance of the spectral estimate.   *(54)*

$w(\mathbf{x})$ Dot noise. The visual noise local to a single pixel in the physical reconstruction function.   *(16,17)*

$\mathbf{x}$ Continuous-space spatial vector, $\mathbf{x} = \begin{bmatrix} x_1 \\ x_2 \end{bmatrix}$.   *(19)*

Z Number of elements in an ordered dither threshold array.   *(52)*

References

(Note: the italicized number on the far right of each entry is the page in this book where a citation was made.)

[1] Adobe Systems (1985)   *(78)*
      Postscript Language Reference Manual.
      Reading, MA: Addison-Wesley.

[2] Allebach, J.P. and B. Liu (1976)   *(265)*
      "Random quasi-periodic halftone process",
      J. Opt. Soc. Am., vol. 66, pp. 909–917.

[3] Allebach, J.P. and B. Liu (1977)   *(41)*
      "Analysis of halftone dot profile and aliasing in the discrete bi-
      nary representation on images",
      J. Opt. Soc. Am., vol. 67, pp. 1147–1154.

[4] Allebach, J.P. (1978)   *(265)*
      "Random nucleated halftone screening",
      Photogr. Sci. Eng., vol. 22, no. 2, pp. 89–91.

[5] Allebach, J.P. (1980)   *(35)*
      "Binary display of images when spot size exceeds step size",
      Applied Optics, vol. 19, pp. 2513–2519.

[6] Allebach, J.P. (1981)   *(6)*
      "Visual model-based algorithms for halftoning images",
      Proc. SPIE, vol. 310, pp. 151–158.

[7] Anastassiou, D. and K.S. Pennington (1982)   *(334)*
      "Digital Halftoning of Images",
      IBM J. Res. Develop., vol. 26, pp. 687–697.

[8] Baldwin, M.W. (1940)   *(190)*
      "The subjective sharpness of simulated television images",
      Proc. IRE, Oct., pp. 458–468.

[9] Bartlett, M.S. (1955)   *(54)*
    An Introduction to Stochastic Processes with Special Reference to
    Methods and Applications.
    New York: Cambridge University Press, pp. 274–284.

[10] Bayer, B.E. (1973)   *(127, 131)*
    "An optimum method for two level rendition of continuous-tone
    pictures",
    Proc. IEEE Int. Conf. Commun., Conference Record, pp. (26-11)–
    (26-15).

[11] Billotet-Hoffman, C. and O. Bryngdahl (1983)   *(265)*
    "On the error diffusion technique for electronic halftoning",
    Proc. SID, vol. 24, pp. 253–258.

[12] Bracewell, R.N. (1978)   *(48, 346)*
    The Fourier Transform and Its Application.
    New York: McGraw-Hill.

[13] Campbell, F.W., J.J. Kulikowski, J. Levinson (1966)   *(84)*
    "The effect of orientation on the visual resolution of gratings",
    J. Physiology London, vol. 187, pp. 427–436.

[14] Chao, Y. (1982)   *(123)*
    "An investigation into the coding of halftone pictures",
    M.I.T. Ph.D. Thesis.

[15] Clapper, F.R. and J.A. Yule (1953)   *(39)*
    "The effect of multiple internal reflections on the densities of
    half-tone prints on paper",
    J. Opt. Soc. of Am., vol. 43, no. 7, pp. 600–603.

[16] Cornsweet, T.N. (1970)   *(79)*
    Visual Perception.
    New York: Academic Press.

[17] Dippe, M.A. and E.H. Wold (1985)   *(265)*
    "Antialiasing through stochastic sampling",
    Computer Graphics (AMC SIGGRPAPH'85 Conf. Proc.), vol. 19,
    no. 3, pp. 69–78.

[18] Dudgeon, D.E. and R.M. Mersereau (1984)   *(44)*
    Multidimensional Digital Signal Processing.
    Englewood Cliffs, NJ: Prentice-Hall, pp. 39–41.

[19] Engeldrum, P.G. (1985)   *(341)*
    "Computing color gamuts of ink-jet printing systems",
    SID Int. Sym. Digest of Tech. Papers, pp. 385–385.

[20] Floyd, R.W., and L. Steinberg (1975)   *(239, 241)*
    "Adaptive algorithm for spatial grey scale",
    SID Int. Sym. Digest of Tech. Papers, pp. 36–37.

[21] Floyd, R.W., and L. Steinberg (1976) *(239, 241)*
     "An adaptive algorithm for spatial greyscale",
     Proc. SID, vol. 17/2, pp. 75–77.

[22] Freund, J.E. (1971) *(56)*
     Mathematical Statistics.
     Englewood, NJ: Prentice-Hall.

[23] Gall, W., K. Wellendorf, and K. Kiel (1985) *(99, 265)*
     "Production of screen printing blocks",
     U.S. Patent 4499489.

[24] Garcia, A. (1986) *(10)*
     "Efficient rendering of synthetic images",
     M.I.T. Ph.D. Thesis.

[25] Gardner, M. (1978) *(233, 338)*
     "White and brown music, fractal curves and one-over-f fluctua-
     tions",
     Scientific Am., Apr., pp. 16–32.

[26] Goodall, W.M. (1951) *(63, 340)*
     "Television by pulse code modulation",
     Bell Sys. Tech. Journal, vol. 30, pp. 33–49.

[27] Graham, C.H., ed. (1965) *(79)*
     Vision and Visual Perception.
     New York: John Wiley & Sons.

[28] Hamill, P. (1977) *(77)*
     "Line printer modification for better grey level pictures",
     Computer Graphics and Image Proc., vol. 6, pp. 485–491.

[29] Hecht, E. and A. Zajac (1974) *(64)*
     Optics.
     Reading, MA: Addison-Wesley, pp. 361–363.

[30] Heckbert, P.S. (1982) *(342)*
     "Color image quantization for frame buffer display",
     Computer Graphics (AMC SIGGRPAPH'82 Conf. Proc.), vol. 16,
     no. 3, pp. 297–307.

[31] Higgins, G.C. and K. Stultz (1948) *(79)*
     "Visual acuity as measured with various orientations of a
     parallel-line test object,"
     J. Opt. Soc. Am., vol. 38, no. 9, pp. 756–758.

[32] Holladay, T.M. (1980) *(99)*
     "An optimum algorithm for halftone generation for displays and
     hard copies",
     Proc. SID, vol. 21, no. 2, pp. 185–192.

[33] Holman, L.A. (1926)   *(71, 78)*
      The Graphic Process, a Series of Actual Prints.
      Boston: Charles E. Goodspeed & Co.

[34] Jaeger, C.W., H. McManus, and D. Titterington (1984)   *(17)*
      "The influence of ink/media interactions on copy quality in ink-
      jet printing",
      Proc. SID, vol. 25/1, pp. 65–70.

[35] Jarvis, J.F., and C.S. Roberts (1976)   *(333)*
      "A new technique for displaying continuous-tone images on a
      bilevel display",
      IEEE Tran. on Commun., vol. COM-24, pp. 891–898.

[36] Jarvis, J.F., C.N. Judice, and W.H. Ninke (1976)   *(6, 241, 253)*
      "A survey of techniques for the display of continuous-tone pic-
      tures on bilevel displays",
      Computer Graphics and Image Processing, vol. 5, pp. 13–40.

[37] Kermisch, D. and P.G. Roetling (1975)   *(41)*
      "Fourier spectrum of halftone images",
      J. Opt. Soc. Am., vol. 65, pp. 716–723.

[38] Knowlton, K. and L. Harmon (1972)   *(77, 157)*
      "Computer-produced greyscales",
      Computer Graphics and Image Proc., vol. 1, pp. 1–20.

[39] Knuth, D.E. (1981)   *(64)*
      The Art of Computer Programming, vol. 2.
      Reading, MA: Addison-Wesley, Ch. 3, "Random Numbers".

[40] Kuhn, L. and R.A. Myers (1979)   *(17)*
      "Ink-jet printing",
      Scientific American, vol. 240, no. 4, pp. 162–178, April.

[41] Lewis, W.G.B. and W.A. Pite, ed. (1902)   *(80)*
      Plate 26: "photo of Eltham Place, Kent",
      The Architectural Association Sketchbook, 3rd series, vol. VI, Lon-
      don: C.F. Kell and Son.

[42] Limb J.O. (1969)   *(127)*
      "Design of dither waveforms for quantized visual signals",
      Bell Sys. Tech. J., Sep., pp. 2555–2582.

[43] Lippel, B. and M. Kurland (1971)   *(127)*
      "The effect of dither on luminance quantization of pictures",
      IEEE Trans. Comm., vol. COM-19, no. 6, pp. 879–888.

[44] Lippel, B. (1976)   *(127)*
      "Two and three-dimensional ordered dither in bi-level picture
      displays",
      Proc. SID, vol. 17/2, pp. 115–121.

[45] Lippel, B. (1978) *(333)*
"Comments on 'A new technique for displaying continuous-tone images on a bilevel display' ",
IEEE Trans. Commun., vol. COM-26, pp. 309–310.

[46] Lowel, G (1916) *(81)*
photo: "Summer house on road north of Florence", *Smaller Italian Villas and Farmhouses.*
New York: The Architectural Book Publishing Co., p. 76.

[47] Mandelbrot, B. and R. Voss (1983) *(338)*
"Why is nature fractal and when should noises be scaling?",
Noise in Physical Systems and 1/f Noise.
New York: North Holland Physics Publishing, pp. 31–39.

[48] Marks, R.J. (1986) *(21)*
"Multidimensional-signal sample dependency at Nyquist densities",
J. Opt. Soc. Am. A, vol. 3, no. 2, pp. 268–273.

[49] Mersereau, R.M. (1978) *(21)*
"Two-dimensional signal processing from hexagonal rasters",
IEEE International Conf. on Acoustics, Speech, and Signal Processing, pp. 739–742.

[50] Mersereau, R.M. (1979) *(21, 44)*
"The processing of hexagonally sampled two-dimensional signals",
Proc. IEEE, vol. 67, no. 6, pp. 930–949.

[51] M.I.T. Museum (c. 1929) *(10)*
Photo: "Building 10", contact print from 5x7 glass negative.

[52] Musha, T. (1981) *(338)*
"1/f fluctuations in biological systems",
Proc. 3rd symposium on 1/f fluctuations, pp. 143–146.

[53] Oppenheim, A.V. and R.W. Schafer (1975) *(54)*
Digital Signal Processing.
New York: Prentice-Hall, pp. 548–549.

[54] Papoulis, A. (1984) *(56)*
Probability, Random Variables, and Stochastic Processes.
New York: McGraw-Hill, p. 178.

[55] Perry, B. and M.L. Mendelsohn (1964) *(77)*
"Picture generation with a standard line printer",
Comm. of the ACM, vol. 7, no. 5, pp. 311–313.

[56] Petersen, D.P. and D. Middleton (1962) *(44)*
"Sampling and reconstruction of wave-number-limited functions in N-dimensional Euclidean spaces",
Information and control, vol. 5, pp. 279–323.

[57] Ratzel, J.N. (1980)   *(343)*
     "The discrete representation of spatially continuous images",
     M.I.T. Ph.D. Thesis.

[58] Roberts, L.G. (1962)   *(3, 63, 340)*
     "Picture coding using pseudo-random noise",
     IRE Trans. Infor. Theory, vol. IT-8, pp. 145–154.

[59] Robinson, A.H. (1973)   *(41)*
     "Multidimensional Fourier transforms and image processing with
     finite scanning apertures",
     Applied Optics, vol. 12, no. 10, pp. 2344–2352.

[60] Roetling, P.G. (1976)   *(334)*
     "Halftone method for edge enhancement and moire suppression",
     J. Opt. Soc. Am., vol. 66, pp. 985–989.

[61] Roetling, P.G. (1976)   *(79)*
     "Visual performance and image coding",
     Proc. SID, vol. 17/2, pp. 111–114.

[62] Roetling, P.G. (1977)   *(6)*
     "Binary approximation of continuous-tone images",
     Photographic Science and Engineering, vol. 21, pp. 60–65.

[63] Roetling, P.G. and T.M. Holladay (1979)   *(35)*
     "Tone reproduction and screen design for pictorial electrographic
     printing",
     J. Applied Photographic Eng., vol. 5, no. 4, pp. 179–182.

[64] Rosenfeld, G. (1984)   *(99, 265)*
     "Screened image reproduction",
     U.S. Patent 4456924.

[65] Schreiber, W.F. (1976)   *(11)*
     "Laser scanning for the graphic arts",
     Proc. SPIE, vol. 84, pp. 21–26.

[66] Schreiber, W.F. (1981)   *(266)*
     "The representation of continuous tone images on binary record-
     ing material",
     Unpublished paper.

[67] Schreiber, W.F. (1983)   *(11)*
     "An electronic process camera",
     Tech. Ass. of the Graphic Arts Proc., May.

[68] Schreiber, W.F. (1983)   *(78)*
     RLE Progress Report No. 125, MIT, p. 6.

[69] Schreiber, W.F. and D.E. Troxel (1985)   *(9, 334)*
     "Transformation between continuous and discrete representation
     of images: a perceptual approach",
     IEEE Trans. PAMI, vol. PAMI-7, no. 2, pp. 178–186.

[70] Schreiber, W.F. (1986)   *(79, 334)*
    Fundamentals of Electronic Imaging Systems: Some Aspects of
    Image Processing.
    New York: Springer-Verlag.

[71] Schroeder, M.R. (1969)   *(341)*
    "Images from computers",
    IEEE Spectrum, vol. 6, 66–78.

[72] Shaw, R., P.D. Burns and J.C. Dainty (1981)   *(17)*
    "Particulate model for halftone noise in electrophotography I.
    Theory, and II. Experimental verification",
    Proc. SPIE, vol. 310, pp. 137–150.

[73] Smithsonian Institution (1844)   *(170)*
    Jacquard-woven silk picture, *"Visite de Msr le Duc d'Aumale a
    la Crois-Rousse, dans l'atelier de M. Carquillat."*
    Photo number 61608.

[74] Sonnenberg, H. (1982)   *(28)*
    "Laser-scanning parameters and latitudes in laser xerography",
    Applied Optics, vol. 21, pp. 1745–1751.

[75] Sonnenberg, H. (1983)   *(28)*
    "Designing Scanners for Laser Printers",
    Lasers & Applications, April, pp. 67–70.

[76] Stevenson, R.L. and G.R. Arce (1985)   *(36, 241, 309)*
    "Binary display of hexagonally sampled continuous-tone im-
    ages",
    J. Opt. Soc. Am. A, vol. 2, no. 7, pp. 1009–1013.

[77] Stoffel, J.C. and J.F. Moreland (1981)   *(6)*
    "A survey of electronic techniques for pictorial reproduction",
    IEEE Tran. Commun., vol. 29, 1898–1925.

[78] Stoffel, J.C. (1982)   *(6)*
    Graphical and Binary Image Processing and Applications.
    Dedham, MA: Artech House.

[79] Stucki, P. (1979)   *(6)*
    "Image processing for document reproduction",
    in *Advances in Digital Image Processing.*
    New York: Plenum Press, pp. 177–218.

[80] Stucki, P. (1981)   *(36, 241, 253)*
    "MECCA -a multiple-error correcting computation algorithm for
    bilevel image hardcopy reproduction",
    Research Report RZ1060, IBM Research Laboratory, Zurich,
    Switzerland.

[81] Taylor, M.M. (1963)   *(79, 185)*
    "Visual discrimination and orientation",
    J. Opt. Soc. Am., vol. 53, June, pp. 763–765.

[82] Ulichney, R.A. and D. Troxel (1982)   *(9)*
     "Scaling Binary Images with the Telescoping Template",
     IEEE Trans. on Pattern Analysis and Machine Intelligence,
     vol. PAMI-4, no. 4, pp. 331–335.

[83] Ulichney, R.A. (1985)   *(127)*
     "Generalized ordered dither",
     M.I.T., ATRP-T-51.
     also Digital Equipment Corporation, DEC-TR-412.

[84] Voss, R.F. and J. Clarke (1975)   *(338)*
     " '1/f noise' in music and speech",
     Nature, vol. 258, no. 5533, Nov. 27, pp. 317–318.

[85] Voss, R.F. and J. Clarke (1978)   *(338)*
     " '1/f noise' in music: music from 1/f noise",
     J. Acoustic Soc. Am., vol. 63, no. 1, pp. 258–263.

[86] Voss, R.F. (1979)   *(338)*
     "1/f (flicker) noise: a brief review",
     Proc. of the 33rd Annual Symposium on Frequency Control,
     May 30–June 1, 1979, pp. 40–46.

[87] White, J.M. (1980)   *(334)*
     "Recent advances in thresholding techniques for facsimile",
     J. Applied Photographic Eng., vol. 6, pp. 49–57.

[88] Witten, I.H. and M. Neal (1982)   *(266)*
     "Using peano curves for bilevel display of continuous-tone im-
     ages",
     IEEE CG&A, May, pp. 47–52.

[89] Woo, B. (1984)   *(266)*
     "A survey of halftoning algorithms and investigation of the error
     diffusion technique",
     M.I.T. S.B. Thesis.

[90] Yule, J.A.C. (1967)   *(79, 99, 341)*
     Principles of Color Reproduction,
     New York: John Wiley & Sons.

Index